Selective Oxidation
of
Hydrocarbons

Selective Oxidation
of
Hydrocarbons

D. J. HUCKNALL
Department of Chemistry,
The City University

1974

ACADEMIC PRESS · LONDON · NEW YORK
A Subsidiary of Harcourt Brace Jovanovich, Publishers

ACADEMIC PRESS INC. (LONDON) LTD.
24/28 Oval Road
London NW1

United States Edition published by
ACADEMIC PRESS INC.
111 Fifth Avenue
New York, New York 10003

Library of Congress Catalog Card Number: 73-19013
ISBN 0-12-358950-9

PRINTED IN GREAT BRITAIN BY
WILLIAM CLOWES AND SONS LTD.
LONDON, COLCHESTER AND BECCLES

Preface

Presently, society is built on a cornerstone of oil and natural gas; most of our energy is derived from this source, as is an increasing proportion of the raw materials required by the chemical industry. For example, the principal chemical feedstock obtained by the processing of crude oil is naphtha which, in Europe, is the prime source of petrochemicals such as ethylene, propylene, butenes, aromatics, etc., and fuels such as motor gasoline and aviation spirit. It seems almost certain that naphtha will remain the predominant feedstock for the petrochemicals industry to the end of this decade and probably beyond and, certainly, as the demand for gasoline continues to increase, our dependence on crude oil is very real.

Unfortunately, as exploitation of crude oil continues to accelerate, it has been realized that the known and proven oil reserves in the non-communist world will be exhausted in about twenty years' time and those unsurveyed areas of the world with geological features favouring deposits of oil and gas are much smaller than those already being worked. These considerations, combined with political factors in certain countries restricting the extraction of crude oil make an "energy crisis", in terms of availability and cost, seem likely.

A number of temporary solutions to the problem have been suggested, usually originating from biased sources, and include the abandonment of legislation affecting the environment, such as that relating to leaded gasolines or the exploitation of protected areas here and in the United States. These proposals, however, by their nature, show how badly misjudged has been the situation. Any measure to avert the "crisis" must, of necessity, suggest ways of bringing about great economies in the use of crude oil. Since the production of gasoline demands between about two and three times the amount of naphtha consumed in the manufacture of petrochemicals, one staggering economy is obvious.

A less spectacular, but equally important source of economy may be found within the petrochemical industry itself and involves increasing the efficiency of existing processes and, perhaps, the use of feedstocks which hitherto had been thought useless. Since most processes within the petrochemicals industry involve catalysis, an excellent starting point for improving process efficiency involves designing the most selective and active catalysts for that process.

The aim of this book has been to examine one such process (the selective oxidation of C_2 to C_5 olefins and alkanes) and critically to review our knowledge thereon. Although it may appear at first glance to be a mere compendium of catalysts, kinetic schemes, etc., it is hoped that the information has been collated in such a way that the search for a new oxidation catalyst will be less empirical. As such, this book is addressed to anyone in either industrial or academic establishments whose work, directly or indirectly, is concerned with the catalytic oxidation of hydrocarbons.

The author wishes to thank Professor C. F. Cullis for the initial suggestion that this monograph should be written. Throughout its preparation, his comments and cogent criticisms were of the greatest value.

Thanks are also due to Mrs R. Hills and, particularly, to Miss D. E. Walsh for typing the manuscript and to Mrs E. A. M. Burton for typing the extremely complex tables of patents.

September, 1973   D. J. HUCKNALL

Contents

CHAPTER 1

Introduction

1.1. General Introduction

Hydrocarbon derivatives containing oxygen and other hetero-atoms are important intermediates in the petrochemical industry, and the future promises further substantial increases in demand. These compounds are much more expensive than the parent hydrocarbons and, particularly in the case of oxygen-containing compounds, a great deal of effort has been expended in attempts to develop processes based upon homogeneous gas-phase oxidation which will produce such products selectively. Unfortunately, homogeneous oxidations are generally unselective, giving a complex mixture of potentially valuable products which would require an elaborate refinery for separation. Thus efforts have been directed towards the production of such compounds by catalytic oxidation, ammoxidation, dehydrogenation, etc. Reactions of this type which are commercially important include:

$$\text{ethylene} + O_2 \rightarrow \text{ethylene oxide}$$
$$\text{ethylene} + O_2 \rightarrow \text{acetaldehyde, acetic acid}$$
$$\text{isobutene, propylene} + O_2 \rightarrow \text{acrolein, methacrolein}$$
$$\text{isobutene, propylene} + NH_3 + O_2 \rightarrow \text{methacrylonitrile, acrylonitrile}$$
$$\text{propylene} + O_2 \rightarrow \text{acetone}$$
$$n\text{-butenes} + O_2 \rightarrow \text{buta-1,3-diene}$$
$$\text{butane} + O_2 \rightarrow \text{maleic anhydride}$$

Fairly extensive reviews of the heterogeneous oxidation of hydrocarbons have been given by Margolis [1] in 1963 and by Sampson and Shooter [2] in 1965. Both reviews attempted to correlate certain properties of the catalysts with their performance in oxidation reactions. Thus, for example, Margolis stressed the importance of the electron work function, whilst Sampson and Shooter qualitatively classified oxide catalysts according to the type of semi-conductivity present. Quite recently too, Voge and Adams have reviewed specifically the catalytic oxidation of lower (C_2 to C_5) olefins [3]; this review was, however, rather limited in its scope. Similarly, Sachtler [4] has produced a short review. The past ten years have seen such activity in the field of catalytic oxidation, ammoxidation and dehydrogenation reactions and their mechan-

1

isms, that there is again a need to collate the published information and re-examine the possibility of producing a unified theory of oxidation catalysis.

After a brief introduction in which the physical and chemical aspects of catalysis are discussed, the present review deals in detail with the oxidation, and closely related reactions, of certain aliphatic hydrocarbons, particularly the C_2 to C_5 olefins although some alkanes are also included. Descriptions are given of the processes and mechanisms involved, not only in the now well-established catalytic oxidations such as ethylene to ethylene oxide, propylene to acrolein and the n-butenes to buta-1,3-diene, but also newer and possibly less well-known processes. Examples of such reactions are the oxidation of ethylene to acetalde-hyde and acetic acid and the oxidation, in the presence of water vapour, of propylene to acetone. Topics are discussed specifically under the process in-volved, for example, the ammoxidation of propylene to acrylonitrile and effec-tive catalysts are discussed thereunder. Certain catalysts (bismuth molybdate, for instance), however, are very versatile and are effective in several different processes. In these cases, the catalysts are discussed in relation to that process, although some overlap is inevitable. The literature available up to the end of 1972 has been covered, with particular emphasis on the period from 1965 to 1972.

1.2. Physical Aspects of Catalysis

Reactions occurring in the presence of a catalyst can be envisaged according to the following processes:

(1) Diffusion of the reactants to the catalyst surface
(2) Adsorption of the reactants on the surface
(3) Reaction on the surface
(4) Desorption of products from the surface
(5) Diffusion of products away from the surface.

The observed rate of a catalytic reaction and the nature of the products formed can be profoundly affected by the rate of mass-transfer, i.e. by how fast reactant molecules are transported to the catalyst surface and by how fast the product molecules are removed. Generally, steps 2, 3 and 4 are considered together, although Beranek [5] has discussed theoretically, anomalies caused by the slow adsorption of the reactant in certain types of consecutive and branched-chain heterogeneous catalytic reactions. However, where steps 2, 3 and 4 are slower than diffusion processes, true "chemical" kinetics will be observed. If the re-verse is true, then "diffusion" kinetics will be observed and the results for selectivity and activation energy will not be characteristic of the chemical reactivity of the system. A monograph by Satterfield and Sherwood [6], exten-sively treated the influence of diffusional limitations in gas-phase reactions. Recently, this work has been revised and the latest monograph [7] describes the theory and computational work which has been done in the field of mass-transfer between reacting fluids and solid catalysts.

The problem of diffusion can be separated into two parts, diffusion or mass-transfer to the external surface of the catalyst, and for those catalysts which are porous, diffusion within the catalyst pores.

The total rate of a chemical reaction on a porous catalyst pellet depends upon the extent of the internal surface and the size of the pores comprising this surface. The influence of intra-particle mass- and heat-transfer has initiated considerable study. Wheeler [8] was apparently the first to recognize that the product distribution obtained in a reactor wherein concurrent and consecutive reactions are taking place, can be influenced by mass-diffusion within the catalyst. Assuming an isothermal catalyst pellet, he distinguished between three types of selectivity, corresponding to the type of reaction occurring e.g. simultaneous first-order reactions, consecutive first-order reactions or parallel first-order reactions. Butt [9] has treated the case of consecutive first-order reactions in non-isothermal pellets and van der Vusse [10], zero- and second-order reactions. Mingle and Smith [11] determined the effects of both concentration- and temperature-gradients on the effectiveness of the internal surface of a catalyst, also taking into account the pore-size distribution within the pellet. Carberry [12] also derived a general expression for isothermal selectivities obtained with porous catalyst pellets, consisting of micropores branched from macropores, for consecutive first-order reactions and Beek [13] obtained a relationship between pellet-size and catalyst performance, again for first-order reactions, but taking into account both temperature- and concentration-gradients. Very recently, Roberts [15] has treated the influence of intra-particle mass diffusion on the overall and relative rates of parallel reactions of the type A $\rightarrow$ B and A $\rightarrow$ C for an isothermal catalyst pellet. Roberts' calculations involved three pairs of reaction orders, (0, 1), (0, 2) and (1, 2), and showed that internal concentration gradients can cause the yield of product formed in the higher-order reaction to decrease by as much as 80%.

Finally, for second-order reactions, under diffusion-controlled conditions, the peculiar values of activation energy and reaction order which can be obtained have been treated by Tartarelli et al. [16–20] and expressions relating catalytic activity and selectivity to the geometrical features of macro-/micro-porous catalysts have recently been derived by Ferraiolo and Beruto [21].

Wicke [22] and also Boreskov [23] have very recently reviewed certain physical aspects of heterogeneous catalysis and useful general summaries of the influence of mass-transfer and pore-diffusion in catalysis, together with daignostic tests, are also to be found in reviews by Shooter [2, 24] and in Levenspiel's book [25].

The diffusion and sorption effects indicated above, require that a complete definition of a catalyst includes a full description of its texture (surface area, pore-size distribution, pore shape, etc.), particularly as factors such as the compacting-pressure and conditions for activating the catalyst, can change the pore structure and hence the catalyst behaviour [26, 27]. Until recently, even elementary data upon surface areas and pore-size distributions were rarely reported in the patent literature, but such data are now being reported with increasing frequency and a better correlation of data will follow as it becomes routine to report such measurements. Many commercial instruments are available for such routine measurements [28], but equipment can often be simply constructed [29–33]. Other physical factors, however, such as surface defects, are less easily studied and, in some cases, this may be the reason why some catalysts reported in the literature are difficult to reproduce.

1.3. Chemical Aspects of Catalysis

There is general agreement that the microscopic reactions which cause the macroscopic phenomenon of heterogeneous catalysis are confined to the surface of the solid catalyst, and even when physical factors such as mass-transfer or diffusion have been either eliminated or ignored, a proper description of heterogeneous catalytic oxidation must treat several difficult problems simultaneously. As stated by Voge [34], these problems include:

(1) The characterization of the solid surface in its reactive state
(2) Identification of the oxygen and other species on the surface and the reactions undergone by each
(3) The steps involved in the reaction path
(4) The structures and energies of the intermediates.

The vast amount of information concerning points (1) to (4) for various conversions will be dealt with in detail later in this review. Suffice it to say that active phases have been identified and studied (see for example, Ref. 35) and the organic surface species in catalysis have been successfully identified in many cases, particularly by i.r. spectroscopy [36–39]. Typical examples involving the identification of reaction intermediates, include a study of the formation of π-allyl complexes on surfaces by Dent and Kokes [40] and Naito et al. [41], using infra-red and microwave spectroscopy; similarly, Whitney and Gay [42] used nmr spectroscopy to study the adsorption of ethylene, propylene and the linear butenes on zinc(II) oxide. Of the possible oxygen species involved in catalytic oxidation, Ishida and Doi [43] observed, using esr spectroscopy, four such species adsorbed on the oxides of V, Cr, Mo and W whilst Shvets and Kazanskii [44] and Krylov, Pariiskii and Spiridonov [45] demonstrated the formation of O_2^- and O^- on silica-, magnesia-, zirconia- and alumina-supported catalysts, containing Ti, V, and Mo ions. Van Reijen [46] has recently reviewed the use of esr spectroscopy in the study of chemisorption and catalysis problems.

By means of isotopic labelling, the steps and intermediates in the reaction path have also been extensively studied (see for example, Refs. 47 and 48). Newer techniques such as low-energy electron diffraction (LEED), Auger Emission Spectroscopy (AES) and electron spectroscopy for chemical analysis (ESCA) have become subjects of great interest recently, on account of their potential value in the determination of the structure of surfaces and in yielding information about the chemical species present on a surface. As such techniques are applied, so our knowledge of catalytic oxidation will increase enormously [49–51].

CHAPTER 2

The Catalytic Oxidation of Ethylene

2.1. To Ethylene Oxide

A. Processes

Epoxides such as ethylene oxide are very valuable industrial organic intermediates and their uses include the manufacture of solvents, plasticizers, explosives, epoxide homopolymers and polyesters [52]. It has long been known that ethylene can be oxidized, very selectively, to ethylene oxide over silver catalysts, the first patent involving this process being issued to the Carbide and Carbons Chemical Company in 1935 [53]. Nowadays, this process has been developed to an extent where there are now at least four commercial processes in operation [54, 55]. These processes usually involve conversion in two stages over fixed-bed, supported silver catalysts, although it is claimed [56] that a three-stage converter requires 20% less reactor volume, 20% less catalyst and gives a higher overall yield of ethylene oxide. Ethylene oxidation is highly exothermic, and though there would appear to be considerable advantages in using fluidized solid catalysts for this reaction in order to achieve greater temperature uniformity within the reactor, no commercial plants of this type seem to be in operation. This may be due to the agglomeration of catalyst particles, as noted by Echigoya and Osberg [57] and Boreskov *et al.* [58], although Wasilewski and Kubica [59] have reported that this problem can be overcome simply, by the addition of talc or graphite to the catalyst. However, despite the apparent advantages of fluidized beds, converters based thereon did not show the expected increase in selectivity towards ethylene oxide, compared with a stationary catalyst.

Most commercial catalysts consist of silver (up to 35%), supported, usually, upon carriers of low surface area (<1 m^2 g^{-1}). The carriers are usually silica, aluminosilicates or silicon carbide, having a porosity of 40–50%, although Erdoelchemie G.m.b.H. [60] claim that by incorporating a "blowing agent" such as lactic acid hydrazide to the catalyst during preparation, an increased porosity is obtained, which increases the selectivity of the conversion and can be used at a much lower temperature than usual (160°C). The catalysts are normally prepared by impregnating a base (suitably pre-prepared and pellet-

6

ized) with a solution of a silver compound such as the lactate or nitrate, containing small quantities (up to 4%) of lactates of elements such as Ba, Al, Ca, Ce, Au or Pt, and drying and activating at 300–400°C to deposit silver metal. The apparent advantage of using silver or other lactates, is that they are less soluble than silver nitrate and will therefore crystallize faster during the drying period. This helps to give a more even silver deposit later, by reducing crystal growth. Silver catalysts have also been prepared by mechanical deposition of active silver metal on alundum spheres [61] or inactive electrolytic silver powders [62]. The latter is an interesting catalyst, converting 32–47% ethylene to ethylene oxide with a selectivity of 66–73% at the fairly low temperature of 205°C. An explanation for the inactivity of electrolytic silver in the oxidation of ethylene, may be obtained from the work of Degeilh [63], who has shown that oxygen is not adsorbed on a freshly-prepared, electropolished single crystal of silver. Mistrik, Ondrus and Gregor [64] suggested yet another method of preparing high porosity, high activity silver catalysts of high heat transfer capacity, by leaching (with sodium hydroxide) an alloy containing 20% Ag + 80% Al, previously fabricated into balls. The technology of preparing ultra-dispersed silver catalysts from silver–aluminium alloys (containing <50 w% Ag) has also been described by Khasin et al. [65] and catalysts having surface areas approaching 40 m^2/g with particle sizes less than 1000 Å diameter may be prepared by such techniques.

Electron microscopy [66, 67], has shown that silver exists as spherical particles on the surface of alumina or silica supports, leaving most of the support surface uncovered, and Vasilevich et al. [66] further studied the activity of a supported silver catalyst for the oxidation of ethylene to ethylene oxide, as a function of the silver content. This work showed that there was indeed an optimum concentration of silver beyond which no advantages were gained. Thus catalysts were examined consisting of 5·8–34·8 w% Ag/porous corundum and it was found that a six-fold increase in the silver-content of the catalyst increased the catalytic activity/unit volume by a factor of only 2–2·3 and actually decreased the rate constant/unit volume of catalyst by a factor of more than three. Adsorption measurements and electron micrographs showed that when the silver content of a catalyst was increased from 5·8 to 24·4 w%, it did not lead to an increased covering of the support with silver, but rather led to an increase in silver particle size. Thus the specific surface of the silver decreased whilst the actual silver surface/unit weight of catalyst remained the same.

Harriott [68] has investigated the influence of the support surface area, upon the selectivity of ethylene oxidation. With an Al$_2$O$_3$ support (surface area, 404–8·2 m^2 g^{-1}), ethylene oxide was appreciably converted (presumably to CO$_2$ and H$_2$O), as the surface area increased and reactions on the support itself increased. Thus, as the ratio, silver surface area to support surface area, is increased, so support reactions increase and selectivity is lessened.

Table 1 summarizes some claims for catalyst configurations and gas-feed compositions found in recent patent literature. Particularly interesting is the fact that organic halogen compounds such as 1,2-dichloroethane, chlorobenzene etc. are often added to the basic feed. Apparently greater selectivity towards ethylene oxide is achieved. Thus, Wasilewski [91], using stationary silver catalysts at 230°C, studied the effects of a number of such organic halogen compounds on the oxidation of ethylene to ethylene oxide. The compounds investigated included ethyl chloride, bromide and iodide, 1,2-dichloroethane, 1,1,2-trichloroethylene, 1,1,1,2,2,2-hexachloroethane and chlorobenzene. It was found that each compound affected the selectivity and degree of conversion, but that the selectivity was most markedly increased by ethyl chloride, hexachloroethane, or 1,2-dichloroethane. Janda and Kluckovsky [92] similarly studied the effects of adding up to 1 ppm of 1,2-dichloroethane to the reactant gas stream. Using a moderated silver catalyst which gave a selectivity of 50% towards ethylene oxide when no additive was present, it was found that the optimum concentration of additive was 0·1–0·2 ppm, when the selectivity increased to 60–65%, although the ethylene conversion fell from 35 to 30%. In concentrations above 1 ppm, it was found that the catalyst was poisoned. Kaliberdo *et al.* [93] have also examined the promotor-action of Cl-containing additives on silver catalysts used, however, in the epoxidation of propylene at 380°C. Using 2-^{14}C-labelled *t*-butyl chloride, it was found that the only oxidation products of the promoter were [14]CO_2 and hydrogen chloride. The greater proportion of the *t*-butyl chloride remained on and modified the silver surface. Other elements such as Se, Te and S have also been found to have a beneficial effect upon activity and selectivity.

Ostrovskii *et al.* [94] made a significant contribution towards our understanding of promoter action. Catalysts were prepared containing labelled Se Te, Cl and S, and from the distribution of the radioactivity, it was concluded that the additives became concentrated on the surface under reaction conditions. However, Se and Te, unlike Cl, were retained by the catalyst. As Sampson and Shooter [2] point out, this may indicate why traces of organic halogen compounds must be continuously added to the reactor feed to maintain the selectivity of the catalyst. Ostrovskii and Temkin [95] and Rudnitskii *et al.* [96, 97] have studied the promoter action of sulphur and selenium respectively. Both conclude that the effect of the additives present in the surface, is to affect the mode of chemisorption and hence the reactivity of oxygen on the silver surface. This, however, will be amplified in the next section.

From Table 1 it can be seen that in some industrial catalyst configurations, alkaline earth compounds such as the salts of barium or calcium, are often added. The mechanism of action of these compounds as additives to silver catalysts has recently been investigated by Spath *et al.* [98, 99] and Kim *et al.* [100]. Spath *et al.* found that when 2–20 mole% of alkaline earth compounds were added to standard, pure silver catalysts, the resulting catalysts, although having greatly increased activity, were less selective towards ethylene oxide formation. In a study of the influence of γ-irradiation (from a ^{60}Co source of 10 kCi placed 33 cm from the sample for 12 h) on catalytic selectivity in the oxidation of ethylene over an industrial (Engelhard) silver catalyst containing 0·28 ppm calcium, Carberry and co-workers [101, 102] observed a significant enhancement in the yield of ethylene oxide on irradiated catalysts. ESCA studies [101, 103] revealed that this increase was mirrored by the appearance of calcium upon the surface, the calcium re-diffusing into the bulk upon reducing and then re-oxidizing the catalyst. Carberry attributed the increased yield of ethylene oxide to the formation of calcium "superoxide" on the surface, a postulate seemingly in agreement with the idea that ethylene oxide is generated via surface O_2^- sites and carbon dioxide by O^- sites.

B. Mechanism

The oxidation of ethylene over silver gives ethylene oxide, carbon oxides and water. There is also some evidence that formaldehyde is formed [104]. The selectivity towards ethylene oxide is remarkably constant over a large range of temperature; thus Kenson and Lapkin [105] found no alteration over the temperature range 175–245°C, whilst Voge [34] found only a small decrease in selectivity (from 75 to 68 %) when the reaction temperature was raised from 190 to 320°C. Voge also noted that the selectivity varied little with conversion, falling merely from 71 to 68 % as the conversion increased from 20 to 80 %. However, as can be seen in Fig. 1, this is not in agreement with the results of Kenson and Lapkin. The limiting selectivity of 75–80 mole C_2H_4O/100 C_2H_4 illustrated in Fig. 1, suggested to Voge [34] that this feature may be explained by a chemical coupling of reactions and in particular, conversion to carbon dioxide may be coupled with the oxidation to ethylene oxide. The mechanism of the catalytic oxidation of ethylene to ethylene oxide in the gas phase, has recently been reviewed [106].

The most widely accepted mechanism for ethylene oxidation over silver, has been that due to Twigg [107]. From a study in a flow system, at temperatures from 200 to 350°C and with short contact times, it was concluded that ethylene oxide and carbon dioxide were formed by a parallel-consecutive mechanism. The rate of carbon dioxide formation was found to be somewhat lower from ethylene oxide than from ethylene. Voge and Adams [3] and Sampson and Shooter [2], also believe that certain aspects of ethylene oxidation can be explained by this scheme of first order reactions, i.e.

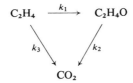

Table 1. Some industrial silver catalyst configurations

| Authors | Comp. of active phase | Carrier | S.A. $(m^2 g^{-1})$ | T°C | Ethylene oxide | | Gas composition | Ref. |
					S %	Yield %		
Engelhard Minerals & Chem. Corp.	Ag	Unspec. inorg. spheres	$Ag = 3$–30	—	—	—	—	[69]
Engelhard Industries Inc.	Ag	Alundum spheres	—	260–270	58–60	35–42	5% C_2H_4 + 6% O_2 + 6% CO_2 + 0·5% C_2H_6 + 2–5 ppm dichloroethane	[70]
Calcagno, Ferlazzo and Ghirga	7–30% Ag (w.r.t. carrier) 0·01–1·0% Pt, Pd or Au (w.r.t. Ag)	α-Al_2O_3	1	265	—	—	6·6% C_2H_4 + 7·7% O_2 + rest N_2 + 3 ppm $C_2H_4Cl_2$	[71]
BASF A.-G.	20 w% Ag, 3·4 w% Ba lactate	α-Al_2O_3	—	235	68	37	—	[72]
Halcon Internat. Inc.	Ag	—	—	—	68·6	—	C_2H_4 + O_2 + 40–60% C_2H_6 + 15 ppm $C_2H_4Cl_2$, CH_2Cl_2, or C_6H_5Cl	[73]
Erdoelchemie G.m.b.H.	Unspec. "porous" Ag + Au, Pt, Ba, Ca or Ce lactates	—	—	160	70–73	—	5 v% C_2H_4 + 88·28 v% air	[60]
Allied Chem. Corp.	18–22% Ag_2O + Ba lactate	α-Al_2O_3 (fired 1650°C)	—	280	—	24·1	4·4% C_2H_4 + 5·4% O_2 + 8% CO_2 (rest N_2)	[74]
Halcon Interna. Inc.	5–25% Ag + Ba lactate	Aluminosilicate spheres (4·8 mm dia.)	<1	280	73	—	5% C_2H_4 + 6% O_2 + 5% CO_2 (rest N_2)	[75]
Farbenfabriken Bayer A.-G. Erdoelchemie G.m.b.H.	35% Ag	α-Al_2O_3 (2–4 μ pores)	—	240–250	—	—	4·5 v% C_2H_4 + 5·5 v% O_2 (rest N_2)	[76]
I.C.I. Ltd.	19 w% Ag + 0·1 w% Ba	SiC (6–8 μ pores)	—	—	71	—	3 v% C_2H_4 + 5·9 v/% (rest N_2) + 0·3 ppm $C_2H_4Cl_2$	[77]
	11·4 w% Ag + 0·12 w% Ba	α-Al_2O_3 (20–180 μ)	—	262	72	—	As above	
Halcon Internat. Inc.	10·88% Ag	Aluminosilicate spheres (≤8 mm dia.)	<1	250	73	—	5% C_2H_4 + 6% O_2 + 5% CO_2 (rest N_2)	[78]
Engelhard Industries Inc.	10% Ag, 0·06% Cu, 0·4% BaO	Alundum spheres	—	230–280	—	34	5% C_2H_4 + 6% O_2 + 6% CO_2 (rest N_2)	[61]
Mistrik, Ondrus & Gregor	20% Ag	Self-supporting	—	226–229	64–71	—	5% C_2H_4 + 7% O_2 (rest N_2)	[64]

	Catalyst	Support / size	Temp	%	Conditions	Ref
Janda and Koma	Ag + 5–10 × 10^{-3} at. % Te Unspec. ceramic	—	240	60	4% C_2H_4 + 6·8% O_2 + 5·8% CO_2 (rest N_2)	[79]
Japan Cat. Chem. Ind. Ltd.	Ag (complexed with alkanolamines) SiO_2–Al_2O_3	—	—	71·3	4·6 v% C_2H_4 + 6·0 v% O_2	[80]
Shell Internationale	3·0 to 10% Ag (dia. = 0·1–0·6 μ) α–Al_2O_3	—	—	72·9	—	[81]
Morikana, Kiyoshi	Acid-washed Ag	—	150–158	75·8†	—	[82]
Dainippon Ink & Chemicals Inc.	Ag + prods. obtained by heating soln. cont. Ag or Mg or $CaCO_3$, $Ba(OH)_2$ and XS $Al_2(SO_4)_3$	—	—	80·0	10% C_2H_4 in air	[83]
Du Pont de Nemours	95%/w Ag + 5%/w Cd as 30-mesh wire	—	255	78	2–4 v% C_2H_4 in air	[84]
Esso Res. and Eng'g. Co.	99·5% Ag + 0·5% Au SiO_2	ca. 30	255	70	1 : 5 : 10 (molar ratio) C_2H_4 : O_2 : He	[85]
Air Products & Chemicals Inc.	Ag aluminate (65 ± 5% Ag)	100–600	190	79·1	2% C_2H_4 + 10% O_2 + 88% inert	[86]
SNAM Progetti S.p.A.	Ag + Ca + Ba Al_2O_3	—	—	76·5	77% C_2H_4 + 5% O_2	[87]
Halcon Internat. Inc.	Ag	—	250	67·8	C_2H_4 (17) + O_2 (7) + N_2 (55) + Ar (15) + C_2H_6 (1) + CO_2 (5v%) + 0·05 vpm $C_2H_4Cl_2$	[88]
Halcon Internat. Inc.	10·88% Ag (+ Ba) Al_2O_3 (3/16 in. diam.)	—	250	73·0	C_2H_4(5%) + O_2(6%) + CO_2(5%) + N_2(84%)	[89]
Battelle Mem. Inst.	Ag + 0·1% Sn alloy‡	—	—	—	—	[90]

† Cf. 48% unwashed.
‡ Catalyst claimed to be 20 times more active than conventional catalysts.

The ratios $k_2/k_1 = 0.08$ and $k_3/k_1 = 0.40$ have been found [3], showing that, as a source of carbon dioxide, the direct oxidation of ethylene is more important than subsequent oxidation of ethylene oxide. The existence of two independent paths for carbon dioxide formation in a parallel-consecutive mechanism, was studied and supported by tracer-work [108], in which ^{14}C-labelled ethylene and inactive ethylene oxide were oxidized. It was found that 80% of the initial carbon dioxide was formed directly from ethylene. It was also shown that when

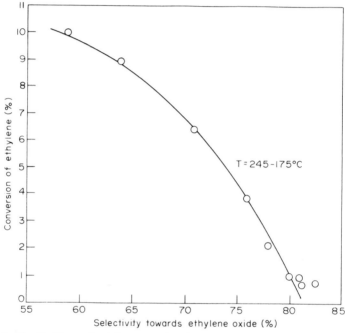

Fig. 1. The limiting selectivity towards ethylene oxide exhibited in the oxidation of ethylene over silver (Ref. Kenson & Lapkin [105].

formaldehyde or acetaldehyde were added to the feed, more ethylene oxide and less carbon dioxide were formed than when ethylene alone was oxidized. The conclusion was then drawn that aldehydes were not important intermediates in carbon dioxide formation. Recent studies [109, 110] have tended to confirm this view.

The problem of the relative importance of the routes to carbon dioxide has always posed difficulties, and conflicting evidence has often been put forward. Thus Hartwig and Bathory reported in one case [111] that kinetic data indicated that carbon dioxide was formed exclusively by the oxidation of ethylene and again [112], that carbon dioxide was formed from ethylene oxide by further oxidation. Since Twigg's original observation [107] that, at temperatures comparable to those used in the silver-catalysed oxidation of ethylene, ethylene oxide isomerized to give acetaldehyde, acetaldehyde has often been suspected as the intermediate in the formation of carbon dioxide from ethylene oxide. Thus, although Margolis and Roginskii [108] had concluded that the isomerization to acetaldehyde, followed by its further oxidation, was too slow to account for the observed CO_2 production, Ide and co-workers [113] in a later investigation of the decomposition and isomerization of ethylene oxide over silver at 210–330°C, confirmed Twigg's observation and concluded that, particularly at low oxygen partial pressures, ethylene oxide was oxidized via acetaldehyde.

The gas-phase pyrolysis of ethylene oxide has also been studied [114, 115]. However, the reaction proposed, viz.

$$CH_2\underset{\diagdown\diagup}{\overset{}{-}}CH_2 \longrightarrow CH_2\overset{\displaystyle \cdot}{-}\overset{\displaystyle \cdot}{CH_2} \longrightarrow CH_3CHO^*$$

would certainly be very slow at the temperatures used for the commercial oxidation of ethylene. On account of the exothermicity of the biradical isomerization, the acetaldehyde produced by such a reaction may have as much as 85 kcal. of excess energy over thermal energy, but experimental observations have been rationalized as a combination of free-radical reactions induced by the decomposition of excited acetaldehyde, and its collisional stabilization. In a very recent study of the homogeneous pyrolysis of ethylene oxide, Rinker et al. [116] concluded that the overall reaction to give, amongst other things, acetaldehyde, was first order with an overall activation energy of 54 kcal. mole^{-1}, and proceeded initially by the reaction

$$CH_2\underset{\diagdown O\diagup}{\overset{}{-}}CH_2 \longrightarrow CH_3CHO^*$$

The activation energy for ring-opening was found to be 52 kcal. mole^{-1} with an A-factor ($2\cdot4 \times 10^{13}$ s^{-1}) agreeing well with A-factors for other unimolecular reactions.

Kenson and Lapkin [105] have recently suggested that adsorbed ethylene oxide would be isomerized much more rapidly than in the gas-phase, if the rate-determining step was its activated chemisorption on the catalyst. Thus a major route to carbon dioxide in the oxidation of ethylene, relative to its direct oxidation, could be isomerization of ethylene oxide to acetaldehyde, followed by oxidation. The mechanism which appeared to fit their kinetic data may be depicted as:

$$H_2C\text{—}CH_2 \quad \xrightleftharpoons{K_a} \quad H_2C\text{—}CH_2 \ (ads.) \tag{1}$$

$$H_2C\text{—}CH_2 \ (ads.) \quad \xrightarrow[\text{slow}]{k_1} \quad H_2C\text{—}CH_2 \ (chemis.) \tag{2}$$

$$\tag{3}$$

(ads. acet-aldehyde)

The overall rate-constant for the latter reaction being k_2. This is followed by:

$$\xrightarrow{k_3} \quad CH_3\text{—}C \tag{4}$$

$$CH_3CHO + 5/2O_2 \quad \xrightarrow{k_4} \quad 2CO_2 + 2H_2O \tag{5}$$

Carbon dioxide formed by direct oxidation of ethylene, may therefore be produced by the reaction sequence

$$\tag{6}$$

the adsorbed ethylene oxide then isomerizing to acetaldehyde which is further oxidized as above. To Kenson and Lapkin, the most logical explanation for ethylene oxide formation, involved the reaction of ethylene with a molecular oxygen–silver complex

$$\tag{7}$$

There would appear, therefore, to be different sites for the oxidation of ethylene to ethylene oxide and for the oxidation of ethylene to carbon dioxide, but with a certain interdependence, and this may be the reason for the limiting selectivity previously mentioned. This site interdependence has also been noted by Ayame and co-workers [117].

The foregoing may be described as a summary of work on the kinetic features of ethylene oxidation over silver and has served to indicate the existence of either two different sites or two different species, which are responsible for ethylene oxidation. Considerable work of a more specific nature has also been carried out, and will be described below. Many studies have been made of the adsorption of various gases on silver. Twigg [107] showed that in the temperature region 200–350°C, chemisorption of ethylene on clean silver was negligible. Later work has confirmed this, and has also shown that the adsorption of carbon dioxide is negligible [118,119]. Ethane, however, is chemisorbed [119], the chemisorption being weak at ambient temperatures but strong at 300°C, the temperature region for the changeover being 150–200°C. Studies with a variety of silver catalysts have, however, confirmed that under conditions of catalysis, only oxygen is significantly chemisorbed [118, 120–122] and oxygen adsorption is undoubtedly critical in explaining the mechanism of ethylene oxidation.

Several fundamental investigations have been made of the adsorption of oxygen on silver. Degeilh [123], for instance, observed the adsorption of oxygen on the (100) and (111) crystallographic planes of an electro-polished silver single crystal from measurements of the contact potential (and hence the work function changes), whilst Bradshaw et al. [124] employed LEED, Auger spectroscopy and surface potential measurements to study adsorption on the (110) plane of a silver single crystal. From their measurements, Bradshaw et al. [124] demonstrated a slow, irreversible adsorption of oxygen on the (110) plane at room temperature, whilst Delgeilh [123] observed the rapid adsorption of oxygen solely on the (100) plane. Despite the obvious value of such studies, perhaps the most useful, with regard to understanding the mechanism of ethylene oxidation, have involved attempts to identify oxygen-containing species on silver under conditions more closely approximating those found in practice. The results of such investigations will now be described.

Thermodynamic data imply, that, under normal catalyst operating conditions, the formation of silver oxides such as Ag_2O, is impossible [125]. Nevertheless, Kagawa et al. [121] examined the surface structure of a silver catalyst (evaporated thin films) during ethylene oxidation, by means of electron diffraction and in situ measurements of i.r. absorption. Two silver oxides (AgO_2 and Ag_2O) were observed upon the catalyst surface, after heating in oxygen. These oxides, however, disappeared after treatment with hydrogen or ethylene. It was supposed, therefore, that under catalytic conditions, ethylene reacts with chemisorbed oxygen or oxygen in the oxide layer to give either carbon dioxide or ethylene oxide Feller-Kniepmeier and co-workers [122] similarly studied technical silver catalysts which had been in service at temperatures in the range 200–300°C. By a combination of X-ray and electron diffraction techniques, they showed that a layer of Ag_2O_3 exists on the surface of the silver crystals and expounded the view that the formation of this oxide was "crucial" for catalysis in this temperature range. Similar stoichiometries have also been reported by May and Linnett [126].

Further to the participation of silver oxides in the epoxidation reaction, Kagawa et al. [127] and Liberti et al. [128] have recently studied the air-oxidation of ethylene in the presence of oxides such as Ag_2O, AgO and Ag_2O_3. In the presence of these oxides, Kagawa et al. [127] found that the rate of formation of ethylene oxide declined in the order:

$$Ag_2O_3 > AgO > Ag_2O$$

By means of electron diffraction, Kagawa et al. [127] had also observed the formation of the oxide AgO_2, and pointed out that the above order coincided with the ease of formation of AgO_2 from these compounds. Indeed, it was postulated that oxygen from the air was chemisorbed on the surface of these oxides with the eventual formation of AgO_2, and that the species O_2^-, present in AgO_2, was responsible for the formation of ethylene oxide.

A comprehensive study of the adsorption and desorption of oxygen on silver powder was made by Czanderna [129]. With the use of a vacuum ultra-micro-balance, the phenomenon was studied in the temperature range −77 to 351°C. Activation energies of 3, 8 and 22 kcal. mole^{-1} were found, and were believed to correspond to the dissociative adsorption of oxygen, molecular adsorption and the mobility of the surface adatoms. The mechanism of adsorption was thought to be:

$$O_2(g) \rightarrow O_2(ads.) \rightarrow O_2^{\delta-}(ads.) \rightarrow 2O^{\delta-}(ads.)$$

and by plotting the experimentally determined activation energies on a graph of θ (proportion of the surface covered by oxygen) against temperature, a diagram was constructed of the various oxygen species existing upon silver at various temperatures and for various values of θ. From this diagram, it would appear that, under normal conditions of catalysis, oxygen exists on a silver surface as either charged atoms (O^-) or charged molecules (O_2^-), a conclusion reached somewhat earlier by Kummer [130]. However, slightly later work, reported by Sandler and Durigon [131] on the desorption and isotopic exchange of oxygen at a porous silver surface, showed that, although chemisorbed oxygen existed in two binding states, no evidence could be found that one of these involved undissociated oxygen. One of the binding states certainly involved strongly-bound oxygen and the results showed that oxygen desportion occurs only from those regions which are free of the firmly-bound oxygen. Sandler continued this work by reporting [132] the effects of "impurities", such as gold and magnesium oxide (known promoters for silver catalysts), on the chemisorption of oxygen on silver. It was concluded that neither was the previously determined value for the activation energy of oxygen desorption (32·5 kcal.) changed by alloy-formation (with gold) or by incorporation of magnesium oxide, nor were the work function or lattice dimensions altered. Impurities, however, eliminated sites for strong oxygen chemisorption and stabilized those planes causing weaker oxygen-adsorption. Recently, Japanese workers have reported some interesting results of great relevance to the nature of the oxygen adsorbed on the surface of silver catalysts. Initially, Ohasi et al. [133] reported that the emission of exo-electrons (very low energy electrons) occurred during oxidation over silver catalysts and that the electron-emission activity was proportional to catalytic activity. Later work by Sato and Seo [134] showed that chemically stimulated exo-electron emission occurs continuously from a silver catalyst during the partial oxidation of ethylene to ethylene oxide. The latest work by Sato and Masahiro [135] has proved that the presence of oxygen in the feed-gases is essential both for exo-electron emission and also for oxidation, and the maximum yield for ethylene oxide coincides with the maximum electron emission. They put forward the theory that, during the partial oxidation of ethylene by silver, the silver surface is simultaneously oxidized by oxygen and reduced by ethylene, i.e. during the reaction a silver oxide containing excess electrons is formed and decomposed in a short time. The excess electrons trapped in the oxide could be emitted if it was assumed that the adsorption of ethylene to form ethylene oxide, lowered the work function of the surface and that continuous emission of electrons would occur as long as the oxide layer was renewed continuously. Other evidence to support this idea is the observation by Sato and Seo [136a, b] that the adsorption of oxygen on silver results in the formation of Ag_2O with the oxygen ion, O_2^-, adsorbed on it, even at temperatures in excess of 190°C, temperatures at which bulk Ag_2O is thermodynamically unstable, together with the fact that ethylene is known to adsorb on oxygenated silver, with donation of electrons. Similarly, Lilov et al. [137] found that the work function of silver may increase or decrease on adsorption of oxygen, depending upon the temperature and pressure of oxygen. The decrease in work function was ascribed to the penetration of the adsorbed oxygen into the surface layers of silver with the probable formation of a non-stoichiometric silver oxide containing defects which are able to trap excess electrons.

Other attempts have been made to solve the problem of whether ethylene oxide formation proceeds via chemisorbed oxygen atoms or via chemisorbed oxygen molecules. For example, Schultze and Teale [138] attempted to determine the products of the reaction of atomic oxygen with ethylene. Oxygen atoms, it was assumed, were produced from decomposition of nitrous oxide adsorbed on silver. Unfortunately, these results may be ambiguous, since Charmon et al. [139] have shown that the surface decomposition of nitrous oxide yields both molecular and atomic oxygen species. Nevertheless, Herzog [140] again utilized the decomposition of nitrous oxide to reinvestigate the problem, and his results are less ambiguous. Nitrous oxide was decomposed at low partial pressures to give a surface on which atomic oxygen was adsorbed; subsequently ethylene was allowed to react on this surface. Carbon dioxide and water were produced almost exclusively and it was also found that the activation energy of oxidation was identical to that for the decomposition of nitrous oxide, the rate-determining step of which can be directly related to dissociative adsorption of this gas. At high nitrous oxide pressures, where formation of molecular oxygen was the preferred reaction of oxygen atoms, the production of ethylene oxide predominated over the production of carbon dioxide and steam. Herzog thus concluded that molecular oxygen (adsorbed) was indeed responsible for ethylene oxide formation.

Table 2. Activation energies for some reactions occurring during the oxidation of ethylene over silver catalysts

Reaction	E_A (kcal.)	Ref.
$C_2H_4 + 3O_2 \rightarrow 2CO_2 + 2H_2O$	23·0	[142]
	7·4	[109]
	15·0	[92]
	21·5	[94]
	21·0	[95]
	12·2	[111]
	29·0 ± 1·7	[105]
$C_2H_4 + \frac{1}{2}O_2 \rightarrow C_2H_4O$	12·3	[111]
	20·6	[62]
	8·45	[109]
	21·4 ± 0·8	[105]
	23·0	[107]
	19·0	[94]
$C_2H_4O + 5/2O_2 \rightarrow 2CO_2 + 2H_2O$	15·2	[111]
	9·8 ± 0·6†	[105]
	11·5	[109]
	18·4	[117]
$C_2H_4O \rightarrow CH_3CHO$	15·0	[113]
	9·8 ± 0·6	[105]
$C_2H_4O \rightarrow C_2H_4 + O$ (ads.)	10·0	[113]

† via acetaldehyde

In conclusion, it should be pointed out that, solely on the basis of the different products (ethylene oxide, carbon dioxide), it may be wrong to infer different types of oxygen present on the surface and, as Sandler and Hickham point out [141], such results can be interpreted even if only one form of oxygen is present. Oxygen molecules, it is further pointed out, as well as atoms, provide a perfect geometrical fit with the two smallest lattice spacings (2·88 and 4·08 Å) of the fcc silver lattice. It may be that the distinction between different types of oxygen may be made on the basis of the way in which the oxygen is embedded into the surface lattice—either as surface or subsurface oxygen.

Table 2 summarizes some recent kinetic data for various reactions occurring in the oxidation of ethylene over silver catalysts. The variation in activation energy illustrated in this table may, perhaps, be due to the presence (either intentionally or unintentionally) of promoters in the catalyst.

2.2. To Acetaldehyde and Acetic Acid

A. Processes

In 1962, Kemball and Patterson reported that ethylene could be oxidized to acetic anhydride and acetic acid over evaporated palladium films [143]. Since then, industrial interest has been aroused in this process and several patents have appeared. Table 3 summarizes some of the recent patents for processes involving conversion of ethylene to acetaldehyde, acetic acid and acetic anhydride. Preferred catalysts appear to consist of palladium (0·06–1·0%) and vanadia (10%), supported on low surface area α-alumina. The temperatures are in the range 140–280°C, contact times are often long (6–11 s) and the oxidant (usually air) is greatly in excess of ethylene.

B. Mechanism

Kemball and Patterson studied the oxidation of ethylene over palladium films in the temperature range 50–140°C and found that the partial oxidation products (acetic anhydride and acetic acid) accounted for only 3% of the total products. Kemball, however, suggested that these products were formed by a path parallel to that for complete oxidation to carbon dioxide. Although none was detected, acetaldehyde was proposed as the intermediate in the partial oxidation, the proposed mechanism being

$$C_2H_4 + O_2 \nearrow CO_2 + H_2O$$
$$\searrow CH_3CHO \xrightarrow{\text{fast}} (CH_3CO)_2O \longrightarrow CH_3CO_2H$$

It was also found that the reaction was, at low pressures, first-order in ethylene and zero-order in oxygen, implying that the surface is covered predominantly with oxygen, ethylene being adsorbed on top, or between, the adsorbed oxygen atoms. At higher pressures, however, the reaction became zero-order in ethylene. Further work by Patterson and Kemball [154] on the oxidation of ethylene over platinum or rhodium showed that acetic anhydride was not a product of this oxidation.

Table 3. Some patented processes for the conversion of ethylene to products other than ethylene oxide

Authors	Comp. of active phase	Carrier	S.A. ($m^2\,g^{-1}$)	T°C (t_c secs.)	Products	Yield (%)	S^y (%)	Gas composition	Ref.
Boutry & Montarnal	0·06% Pd + 10% V_2O_5[1,3]	α–Al_2O_3	0·3	250 (6)	Acetic acid	66	—	—	[144]
	As above[1,3]				CO_2	—	—	—	
		α–Al_2O_3		280	CO_2	—	—	—	[144]
Inst. Franc. du Petrole	As above[1,4]	α–Al_2O_3	0·3	250	Acetic acid	37	—	1% C_2H_4 + 36% air + 63% steam	[144]
		α–Al_2O_3		250	Acetic acid	51·6	—		[145]
	As above + $FeCl_2$	α–Al_2O_3	0·3	250	Acetic acid	59·8	—	—	[145]
	As above + KCl	α–Al_2O_3	0·3	250	Acetic acid	59·0	—	—	[145]
	As above + MoO_3	α–Al_2O_3	0·3	250	Acetic acid	58·5	—	—	[145]
	1 = activated in air (400°C) 2 = nitric acid for impreg. 3 = HCl used for impreg. 4 = 5% H_2SO_4/HCl for impreg.								
Boutry & Montarnal	1% Pd + V_2O_5	—	—	248 (6)	Acetic acid	74	—	4 v% C_2H_4 + 28 v% O_2 + 68 v% steam	[146]
					CO_2	26	—		
B.P. Chemicals (U.K.)	Pd + Cr	Al_2O_3 (fired at 1060°C)	—	180 (10)	Acetic acid	—	60	1 v% C_2H_4 + 32 v% air + 67 v% steam	[147]
					Acetaldehyde	—	28		
					Vinyl acetate	—	2		
					CO_2	—	10		
Union Carbide Corp.	0·8% Pd + 10% V_2O_5	α–Al_2O_3	—	140 (11)	Acetaldehyde	69	—	C_2H_4 : air : steam = 1 : 20 : 9	[148]
					CO_2				
Gerberich & Hall	Pd (metal)	—	—	—	Acetaldehyde	—	—	—	[149]
					Acetic acid	—	—		
					Acetic anhydride	—	—		
Shell Oil Co.	1 atm. Pd^{2+} + 3–15 atm. Cu^{2+} + 3–15 atm. Ti^{3+} + 1–7 atm. alkali metal	Silica gel	—	236	Acetaldehyde	63·6	—	C_2H_4 : O_2 : N_2 : H_2O 1 : 1 : 6 : 6	[150]
					Acetic acid	18·4	—		
					CO_2	13·4	—		
Nat. Distillers and Chem. Corp.	2% Pd, 82% H_3PO_4	SiO_2 or C	—	175	Acetic acid	—	79	—	[151]
Japan Cat. Chem. Ind. Co. Ltd.	V + Pd + Sb	Fused Al_2O_3	—	250 (7·2)	Acetic acid	84·5	—	C_2H_4 : O_2 : H_2O : N_2 2·1 : 21·0 : 45·0 : 31·9	[152]
					Formic acid	1·5	—		
Inst. Franc. du Petrole	0·003% Pd, 0·07% Li, 14·5% U_2O_5	SiC (85·4%)	—	—	Acetaldehyde	55·6	—	—	[153]
					Acetic acid	31·5	—		

A later study by Gerberich and Hall [155], employing a palladium sponge catalyst (0.2 m^2 g^{-1}), in a circulating flow reactor, at temperatures in the range 71–102°C, revealed that the selectivity towards partial oxidation products could be raised from 3 to 30%. The reason for the previous low yield [143] was probably that, in a static system, the surface became progressively poisoned with acetic acid. Gerberich and Hall also noted that the selectivity was almost independent of temperature (33% at 109°C and 38% at 78°C), indicating an activation energy difference between the parallel paths suggested by Kemball, of less than 2 kcal. mole^{-1}. Selectivity increased, however, with increasing partial pressure of oxygen in the system.

Gerberich and Hall [156] have continued to study the oxidation of ethylene over palladium (and extended the work to include palladium-gold alloys). Results were reported for a steady-state reactor and also in a recirculating reactor where the products were stripped at each pass. Over palladium (0.2 m^2 g^{-1}) the partial oxidation products were as previously reported [155], but traces of ethylene oxide were also found. The initial selectivity towards partial oxidation products was between 25 and 45% depending on the reaction conditions, and the steady state selectivity was between 6 and 25%. From initial rate data, the activation energy for both partial and total oxidation of ethylene was found to be 20 ± 2 kcal. mole^{-1}, but under steady-state conditions, this value was 30 ± 2 kcal. mole^{-1}. The latter value may reflect catalyst poisoning by acetic acid and acetic anhydride. In experiments with 1,2-^{14}C-labelled acetaldehyde it was also found that 80% of the activity was recovered in acetic acid and 10% in carbon dioxide. ^{14}C-labelled ethylene oxide was recovered unchanged. This tended to confirm the original mechanism proposed by Kemball and Patterson.

Cant and Hall [157] have recently extended their investigations to a study of the metals Pt, Pd, Ir, Ru and Rh supported upon high surface-area silica. In all cases, ethylene was oxidized to acetaldehyde and acetic acid, but only with palladium and iridium were the quantities of these products significant. The selectivities to partial oxidation products over these metals were:

Metal	Acetaldehyde	Acetic acid
Ru	<0·1	<0·5
Rh	<1	<1
Pd	<3	15–25
Ir	<0·1	8–13
Pt	0·2	0·8

Specific activities lay in the order: Pt, Pd, Ir, Ru, Rn and correlated well with %d-character and atomic radius of the metal; such correlations are, however, of dubious value.

Other studies of noble metal catalysts have been made by Moss and Thomas

Table 4. Kinetic parameters for the total oxidation of ethylene over noble metals

Metal and support	E_A (kcal.)	log A	Ref.
11% Pd–Ag	16·4	(4·19)†	Moss and Thomas [158]
26% Pd–Ag	(18·5)	(5·29)†	
43·5% Pd–Ag	(13·9)	(3·21)†	
62% Pd–Ag	30·2	10·01 †	
Pd(m) unsupp.	13·6	4·03 †	
Au(m) unsupp.	26·0 ± 4·0		Woodward, Lindgren and Corcoran [142]
Pd(m)	14·3		Kemball and Patterson [143]
Pd(m) unsupp.	30 ± 1·7	29·9 ± 1·0‡	Geberich, Cant and Hall [156]
50% Pd–Au	25·6 ± 2·6	25·1 ± 1·4‡	
40% Pd–Au	20·2 ± 1·0	21·7 ± 0·5‡	
20% Pd–Au	21·2 ± 1·8	21·6 ± 0·8‡	
15% Ru–Silica (high S.A.)	25·8 ± 1·4	24·9 ± 0·7‡	Cant and Hall [153]
Rh–	25·4 ± 1·1	23·8 ± 0·5‡	
Pd–	20·4 ± 0·9	24·4 ± 0·5‡	
Ir–	16·9 ± 0·3	21·1 ± 0·2‡	
Pt–	18·5 ± 0·5	23·1 ± 0·5‡	
1·5% Rh/α–Al$_2$O$_3$ (0·3 m^2 g^{-1})	26·5 ± 0·2	—	Cant and Hall [161]
1·5% Ru/α–Al$_2$O$_3$ (0·3 m^2 g^{-1})	29·0 ± 0·4	—	

† In torr CO_2 min^{-1} cm^{-2}.
‡ In ethylene molecules s^{-1} cm^{-2}.

[158] who studied the oxidation of ethylene over evaporated palladium–silver alloy films.

Japanese workers have also investigated the oxidation of ethylene to acetaldehyde and acetic acid. Mitsutani et al. [159] oxidized an ethylene/oxygen/steam mixture over a catalyst of palladous and cuprous chlorides supported on active carbon at 95–100°C and Ishii and Matsuura [160] oxidized ethylene over alumina–molybdena catalysts in the temperature range 250–400°C. In the latter case, the oxidation products were mainly acetic acid, carbon dioxide and water with traces of acetaldehyde. The yields of acetic acid apparently increased as the Mo^{5+} content of the catalyst increased. Results obtained on acetaldehyde oxidation also indicated that acetic acid was formed directly from ethylene.

Finally, Table 4 lists experimentally-determined kinetic parameters for the complete oxidation of ethylene over various noble metals.

2.3. To Other Products

In a recently published paper, Costa Novella et al. [162] have reported the synthesis of glyoxal by the gas-phase oxidation of ethylene at 140°C. Quite high yields (50·1 %; selectivity = 100 %) of glyoxal were obtained over a selenium dioxide–silica catalyst.

CHAPTER 3

The Catalytic Oxidation of Propylene

3.1. To Propylene Oxide

Like ethylene oxide, propylene oxide is an important industrial intermediate and its manufacture by the direct oxidation of propylene is highly desirable. Recently, Monsanto established a small-scale plant to carry out such a reaction, and yields up to 60 % of a propylene oxide/acetic acid mixture are reported. The process is, however, non-catalytic.

Until very recently, attempts to find active and selective catalysts for the vapour-phase epoxidation of propylene met with little success. Silver, the catalyst so effective for ethylene epoxidation, was reported to yield only carbon dioxide and water with propylene [163], although Kaliberdo et al. [164, 165] have reported the formation of significant quantities of propylene oxide under certain conditions. Typically, Kaliberdo et al. [164] oxidized propylene–propane (1:4·5) mixtures over catalysts containing silver metal alone (catalyst (i)) and silver metal + silver(I) oxide (catalyst (ii)). On catalyst (i), 34 % propylene was reportedly converted to oxygen-containing products, comprising 4·3 % propylene oxide (selectivity 14 %), 5·5 % acetaldehyde, 1·6 % acetone, 1·4 % methanol and a little (<0·4 %) propionaldehyde. Catalyst (ii) was less useful, converting a mere 8 % of the propylene to oxygenated compounds. It was thought [165] that the low yields of propylene oxide on silver catalysts, are due to isomerization and decomposition.

Despite the impracticability of developing a manufacturing process based on the silver-catalyzed oxidation of propylene, it has, nevertheless, been thought worthwhile to patent such a process [166, 167]. The use of other supported metals such as rhodium and ruthenium etc. has also been investigated, but the products are numerous, and include acrolein, acetone, C_3 acids, acetaldehyde and acetic acid [157, 161].

In 1972, encouraging claims were made for the catalytic gas-phase oxidation of propylene to propylene oxide. For example, Honda et al. [168] claim its manufacture "in good yield" by the oxidation of propylene, at 200°C, in the presence of thallium oxide, whilst Centola et al. [169] reported epoxidation in the presence of tungsten(VI) oxide and the tungstates of Cu^{2+}, Ag^+, Cd^{2+}, Ni^{2+}, Mn^{2+}, Co^{2+}, Pb^{2+}, Tl^{3+}, Fe^{3+}, and Al^{3+}. In the latter work, although partial oxidation products also included acetone, acetaldehyde, and smaller quantities of formaldehyde, methanol, acetic acid and isopropanol, thallium and iron tungstates yielded propylene oxide with a selectivity of 37% at propylene conversions of 40%.

The most satisfactory process so-far developed for propylene epoxidation, is that developed by Halcon International Inc. and based upon Halcon's original discovery that, in the presence of selected catalysts, hydroperoxides such as ethylbenzene hydroperoxide, can be made to react with olefins in the liquid-phase, to give high yields of both epoxides and alcohols. In the case of propylene, for example, the selectivity towards propylene oxide can be better than 95%. This reaction involved is as follows

$$RO_2H + CH_3{-}CH{=}CH_2 \longrightarrow CH_3{-}\overset{\displaystyle O}{\overset{\diagup\diagdown}{CH{-}CH_2}} + ROH$$

A variety of catalysts can be used, including compounds of molybdenum, vanadium, titanium, niobium, tantalum, tungsten, rhenium, etc. Usually, readily hydrocarbon-soluble forms of the catalyst, such as naphthenates, acetyl acetonates, etc. are preferred, but in the case of molybdenum, less-soluble forms, such as the trioxide, disulphide, etc., have been found highly effective. The kinetics of such a reaction have recently been studied [170, 171, 224], whilst Hiatt [172] and Metelitsa [173] have reviewed the epoxidation of olefins, in general, by hydroperoxides.

3.2. To Acrolein

The discovery, by Hearne and Adams [174], that propylene could be selectively oxidized to acrolein over cuprous oxide, marked the beginning of the now world-wide process of catalytic oxidation of olefins to aldehydes over metal oxide catalysts. However, the present era of olefin oxidation really gathered impetus with the discovery of bismuth molybdate as an oxidation catalyst [175]. The conversion of propylene to acrolein using cuprous oxide has been amply reviewed [1, 2, 3], but work is still continuing on certain aspects and some recent studies will be mentioned here.

A. Cuprous Oxide Catalysts

Evidence has suggested that, over copper catalysts, the oxidation of propylene to acrolein proceeds via a π-bonded allyl surface intermediate [176, 177]. The idea was supported by Zhdanova and co-workers [178] who used i.r. spectroscopy to study the chemisorption of propylene on cuprous oxide supported on high surface area (200 m^2 g^{-1}) alumina. Two types of surface intermediate were detected, one with absorptions corresponding to those of copper(I) acetate

$$
\begin{array}{c}
\quad\quad\quad {}^{CH_3} \\
\quad\quad\quad \diagup \\
O\cdots CH \\
\quad\quad\; \vdots \\
Cu\cdots O
\end{array}
$$

which decomposed to give carbon dioxide, and another, of the type

$$
H_2C \cdots CH \cdots CH_2 \\
\diagdown \quad \diagup \\
Cu
$$

which then gave acrolein.

Other studies have been made with the aim of identifying the active phase in selective copper oxide catalysts. For example, Ghorokhovatskii et al. [179] showed that selective catalysts were composed mainly of copper(I) oxide or copper(I) oxide plus copper metal. Further studies to correlate the catalytic and solid-state properties of copper oxides, were carried out by Wood et al. [180]. Investigations were made into the composition of the solid phase of effective copper catalysts under reaction conditions, by simultaneous in situ measurements of the electrical conductivity and optical properties of a single copper(I) oxide crystal and the kinetics of acrolein formation at 350°C in a differential flow-reactor. It was found that changes in the feed-gas composition caused great changes in the conductivity of the crystal, suggesting that only a relatively thin surface layer participates in the reaction. Results further indicated that the most active solid phase for acrolein formation is stoichiometric- or copper-rich copper(I) oxide. Oxygen-rich copper(I) oxide or copper(II) oxide favour the complete oxidation of propylene. The results were interpreted in terms of the densities of different oxygen species present on the catalyst surface. These species were $O(s)$, $O^-(s)$ and $O^{2-}(l)$ (where (s) and (l) refer to surface and lattice species), supposedly formed from gas-phase oxygen by the following sequence of events

$$\tfrac{1}{2}O_2(g) \;\rightleftharpoons\; O(s) \tag{1}$$

$$O(s) \;\rightleftharpoons\; O^-(s) + p \tag{2}$$

$$O^-(s) \;\rightleftharpoons\; O^{2-}(l) + p + V_{Cu} \tag{3}$$

where p represents a hole, the formation of which is associated with the formation of the oxygen species, and V_{Cu} represents a copper vacancy. The relative distribution of the oxygen species was said to depend, not only on oxygen pressure, but also on the defect structure of the catalyst; thus in copper-rich copper(I) oxide, the lattice has a high affinity for oxygen and the equilibrium in equation (3) lies far to the right. As the lattice becomes oxygen-rich, high hole-densities on the surface result in a shift to the left in equation (3), resulting in the predominance of O(s) on the catalyst.

Recently, Holbrook and Wise [181] have reported an extension of this work to include an investigation of the promoter-action of methyl bromide in propylene oxidation. The enhancement of the specificity of propylene oxidation over copper by gaseous additives such as halogens, halogenated compounds and selenium compounds, has long been known [2], but the results of Holbrook and Wise are particularly interesting. Table 5 shows the profound effect of methyl bromide on the oxidation of propylene over a copper(I) oxide crystal at 350°C at varying oxygen partial pressures.

Table 5. Propylene oxidation at 350°C

	[acrolein]/[CO + CO$_2$]	
P_{O_2} (torr)	With CH$_3$Br†	Without CH$_3$Br
25	0·37	0·042
50	0·38	0·035
65	0·40	0·027
100	0·39	0·011
125	0·37	0·005
150	0·34	—

† [CH$_3$Br]/[C$_3$H$_6$] = 4·5 × 10^{-4}

An explanation of the action of methyl bromide was proposed, based upon a previous model [180], in which this compound anchored the Fermi level of the catalyst even when the concentrations of propylene and oxygen were altered. The process proposed was

$$O^{2-}(s) + Br(s) + p \rightleftharpoons BrO^-(s)$$

At very high concentrations of methyl bromide, the conductivity of the catalyst increased, due, it was proposed, to the reaction

$$Br(s) \rightleftharpoons Br^-(s) + p$$

and under these conditions, a complete loss of catalytic activity was noted.

B. Bi$_2$O$_3$–MoO$_3$ Catalyst

The combination of bismuth oxide and molybdenum oxide is the basic component of the now-famous Sohio catalyst, the development of which has been amply documented [175, 182]. Bismuth molybdate catalysts display extremely high activity and produce acrolein with great selectivity, whether or not oxygen is present. Other products, such as carbon dioxide, carbon monoxide, acetaldehyde, formaldehyde and smaller quantities of propionaldehyde, propionic, formic and acetic acids are also formed [183], and their origin will be discussed in detail later.

(a) *Kinetics.* Studies of the gross kinetic features have shown that, over bismuth molybdate catalysts, the oxidation of propylene to acrolein is first order with respect to the olefin and independent of oxygen and the products. Adams and co-workers [184] found that the conversion-selectivity data could be amply described by the following reaction sequence:

$$CH_2{=}CH{-}CH_3 \xrightarrow{\ k_1\ } CH_2{=}CH{-}CHO$$

$$\underset{k_3}{\searrow} \quad CO_2 \quad \underset{k_2}{\swarrow}$$

At 460°C, $k_3/k_1 = 0.10$ and $k_2/k_1 = 0.25$. Activation energies have also been obtained from the formation of certain products, and these are tabulated below:

Table 6. Activation energies (kcal. mole^{-1}) for the formation of certain products during propylene oxidation over bismuth molybdate catalysts

Product				
Acrolein	Acetaldehyde	CO	CO$_2$	Ref.
31 ± 4^a				[185]
17 ± 4^b				[185]
13	27	24	36	[183]
18†^{c}			15†^{c}	[186]
25†^{d}			23†^{d}	[186]
17·4				[187]
29 ± 5				[188]
20 ± 1				[189]

^{a} No gaseous oxygen present.
^{b} Gaseous oxygen present.
^{c} Bi molybdate/silica gel.
^{d} Bi molybdate/corundum.
† P-promoted.

(b) *Mechanism.* It is now clearly established, by the use of isotopic tracers, that the oxidation of propylene over bismuth molybdate proceeds via the initial formation of the symmetrical allylic intermediate. For example, McCain et al. [190] showed that, in the oxidation of 1-^{13}C-prop-1-ene, both ends of the molecule could be converted to the carbonyl group of acrolein with equal probability. Isomerization of the propylene prior to oxidation was shown to be relatively unimportant, so that the first stage in the oxidation must involve the reaction

$$CH_2{=}CH{-}CH_3 \xrightarrow{\ -H\ } CH_2 \cdots CH \cdots CH_2$$

Once the adsorbed allylic species is formed, its subsequent reactions are less clear and the stages leading to the later incorporation of oxygen are subject to dispute. Early studies by Adams and Jennings [177, 191], using deuterated propylenes, strongly suggested that the allylic intermediate underwent a second hydrogen-abstraction before the incorporation of an oxygen atom to give acrolein. The evidence for this assertion may be presented by considering their scheme [177] for the oxidation of 3-d_1-prop-1-ene

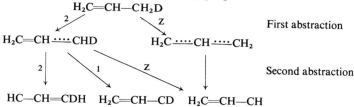

A deuterium atom is more difficult to remove from a molecule than a hydrogen atom, and, at temperatures in the region of 460°C, the rate-constant ratio, k_D/k_H, is about 0·55. If the chance of removing a hydrogen atom is 1, then that of removing deuterium is 0·55(Z), and so the rates of formation of the intermediates are as shown. Attention is drawn particularly to the further reactions of the allyl species. If the incorporation of oxygen occurs immediately after its formation, each end of the intermediate will react with equal probability. However, if abstraction of a further hydrogen atom occurs before incorporation, the deuterated carbon will react at only 75 % of the rate of the undeuterated carbon, and this was shown to be the case experimentally [191, 192]. These results are convincing, but very recently evidence has been obtained [170, 193] that, at least over rhodium and ruthenium catalysts supported on low surface area α-alumina, another mechanism may be operating, in which oxygen is added before the loss of the second hydrogen atom. Evidence has also been obtained by Daniel and Keulks [194] for a similar mechanism operating with bismuth molybdate catalysts. Much of the evidence is based on the observations of certain homogeneous gas-phase reactions occurring within the reactor post-catalyst zone and this will now be dealt with in some detail.

The nature of the minor products of the oxidation of propylene, e.g. acetaldehyde, propylene oxide, propionaldehyde, etc., has led to speculation about their origin. McCain and Godin [195] elegantly demonstrated the simultaneous occurrence of homogeneous and heterogeneous reactions in the oxidation of propylene over a BiMoP catalyst at 445°C. Using a reactor designed so that the empty volume which followed the catalyst could be varied, they found that if the residence time in this volume, of the exit gases from the catalyst, was increased, there was a corresponding increase in the yields of acetaldehyde and propylene oxide. It was concluded that propylene oxide was formed exclusively in this after-zone, whilst acetaldehyde formation occurred both in and after the catalyst bed. Keulks *et al.* [188] also reported that, during the partial oxidation of propylene to acrolein over bismuth molybdate, acrolein underwent a number of homogeneous reactions in the post-catalyst zone. These reactions were dependent on the partial pressure of oxygen and the volume of the post-catalyst zone. It also appeared that the majority of side-products could be explained in terms of the further oxidation of acrolein.

In a very recent paper, Daniel and Keulks [194] reported that, at 450°C, an enhanced conversion of propylene occurred in the presence of bismuth molybdate, in a reactor having a large post-catalyst volume. However, without bismuth molybdate, no reaction occurred. In such a reactor, a variety of products was produced, and some of these, such as propylene oxide, propionaldehyde, methane and formaldehyde, were not produced in exclusively heterogeneous reactions. It appeared that a surface-initiated homogeneous gas-phase reaction had occurred. The ultimate conclusion was that, as previously suggested by Margolis [196], the sequence of events was initiated by free allyl radicals. Indeed, Hart and Friedli [197] have presented direct evidence for the desorption of allyl radicals in the oxidation of propylene over manganese oxide at temperatures above 525°C, the surface probably acting as a hydrogen-abstracting agent. Once the free allyl radical has been formed, it can either react with gas-phase oxygen to form free allyl peroxide or hydroperoxide or, (and this was thought more likely), react with oxygen bound to the surface to form a surface allyl peroxide or hydroperoxide. The latter was then desorbed into the gas phase and acted as chain-initiator, reacting further with propylene to yield propylene oxide or decomposing at the surface to give acrolein. The proposed reaction scheme may be summarized:

$$CH_3—CH{=}CH_2(g)$$

$$\Big\downarrow {-H}$$

$$CH_2 {\cdots} CH {\cdots} CH_2\,(ads.) \longrightarrow CH_2{=}CH—CH_2^{\cdot}(g)$$

$$\begin{array}{c} (1)\ {-H} \\ (2)\ {+O(lattice)} \end{array} \qquad {+O_2} \qquad {+O_2}$$

$$CH_2{=}CH—CHO \xleftarrow[{-H_2O}]{\text{surf. reaction}} CH_2{=}CH—CH_2—O—O(H)\ (ads.\ or\ (g))$$

$$\text{homo.}\Big| {+C_3H_6(g)}$$

$$CH_3—\underset{\underset{O}{\diagdown\diagup}}{CH}—CH_2$$

It was also thought that surface-initiated homogeneous reactions may become important at high propylene to oxygen ratios, in reactors having large post-catalyst volumes or with catalysts containing a large fraction of voids; the production of propylene oxide, it was suggested, is indicative of a surface-initiated homogeneous reaction. In this context, Wragg *et al.* [198] detected an "impurity" with mass 58 during the oxidation of propylene, in the absence of oxygen, over bismuth molybdate. This was tentatively assigned to either propylene oxide or acetone, possibly formed in the gaseous phase.

(*c*) *Competitive and consecutive reactions.* As previously pointed out, apart from acrolein and water, the oxidation of propylene over bismuth molybdate yields carbon monoxide, carbon dioxide, saturated aldehydes and acids. A complete picture of the mechanism must, therefore, offer some explanation for the formation of these compounds and much information has been obtained by Russian workers using ^{14}C-labelled propylene, acrolein and acetaldehyde [183, 199, 200]. For example, 1- and 2-^{14}C-prop-1-ene have been oxidized [199, 183] and the distribution of radioactivity in the products, has been determined. Some of these results are shown in Table 7. In the case of 1-^{14}C-prop-1-ene, the

Table 7

Initial compound	Contact time (s)	Molar radioactivity (α)				
		Acrolein	AA†	CH_2O	CO	CO_2
$CH_2=^{14}CH—CH_3$	1·0	100	89	17	58	68
	2·0	100	83	10	42	65
	3·5	100	87	8	51	57
$^{14}CH_2=CH—CH_3$	0·5	100	52	59	56	53
	1·0	100	54	56	65	47

† AA = acetaldehyde.

almost equal distribution of activity between acetaldehyde and formaldehyde indicates that these products are formed by oxidation of an allyl species rather than by oxidative attack at a localized double bond. Gorshkov *et al.* [183] have also studied the kinetics of oxidation of a variety of aldehydes and acids over bismuth molybdate, and the rate coefficients have been determined (Table 8).

Work was also carried out on the oxidation of acrolein labelled in various positions. The observed distribution of radioactivity in the oxidation products can be seen in Table 9. From these results it appears that formaldehyde is derived mainly from the carbonyl group of acrolein and that the oxides of carbon are formed essentially from the vinyl group. Similar experiments with acetaldehyde indicated that formaldehyde and carbon oxides arise from the methyl- and carbonyl-groups respectively. Finally, it should be mentioned that

Table 8

Compound	Rate coeffs. (s^{-1} $mole^{-2}$)		
	CO	CO_2	CH_2O
CH_2O	0·1	0·15	—
CH_3CHO	2·0	1·3	1·8
$CH_2\!=\!CH\!-\!CHO$	0·08	0·13	0·05
HCO_2H	0·4	11·8	—
CH_3CO_2H	2·6	7·4	6·8
$CH_2\!=\!CH\!-\!CO_2H$	0·1	0·5	—

Acrolein oxidation

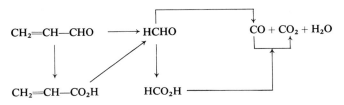

Acetaldehyde oxidation

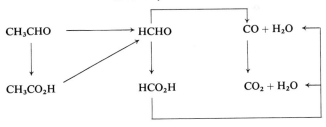

Table 9

	Reaction products Molar radioactivity (α)		
Initial compound	CH_2O	CO	CO_2
$CH_2\!=\!^{14}CH\!-\!CHO$	5	40	38
$^{14}CH_2\!=\!^{14}CH\!-\!CHO$	15	37	40
$^{14}CH_2\!=\!CH\!-\!^{14}CHO$	46	46	44

Keulks and Rosynek [188, 201], using similar techniques, oxidized ^{14}C-labelled- and unlabelled-acrolein and concluded that the carbon dioxide formed in propylene oxidation arises almost exclusively from the further oxidation of acrolein and not from the direct oxidation of propylene, a conclusion conflicting with that of Lapidus *et al.* [202].

(*d*) *The nature of the active phase.* Although the work so far described has been performed on systems containing bismuth molybdate catalysts, it has not emphasized the superior activity, selectivity and versatility of Bi_2O_3–MoO_3 mixtures. Thus, not only is this catalyst excellent for the oxidation of propylene to acrolein, it is also very good for reactions such as the ammoxidation of propylene to acrylonitrile and the oxidative dehydrogenation of the butenes to buta-1,3-diene. As a catalyst, Bi_2O_3 itself has fairly low activity and oxidation in its presence is almost complete. Pure molybdenum oxide has an even lower activity but is fairly selective. In combination, however, the remarkably active and selective bismuth molybdate catalyst emerges, as the results of Lazukin *et al.* [203] indicate. These results compare the activity (expressed as moles propylene oxidized m^{-2} s^{-1}) and selectivity (expressed as mole% propylene converted to acrolein) of such catalysts in the air oxidation of propylene at 450°C.

Catalyst	Activity ($\times 10^7$)	Selectivity
Bi_2O_3	1·65	0
MoO_3	0·46	87
$Bi_2O_3 \cdot 2MoO_3$	21·64	94

Although industrial catalysts contain silica and phosphate as well as Bi_2O_3 and MoO_3, many fundamental studies have employed systems containing only bismuth and molybdenum oxides, in order to throw light on the fundamental issues of whether the observed selectivity and activity are due to an actual compound of bismuth and molybdenum or whether bismuth is merely a very efficient promoter for MoO_3. As a result of such studies it is now known that bismuth molybdate catalysts display their superior properties only if the catalyst composition lies within the composition range Bi/Mo 2/3 to 2/1 (atomic ratio). Attempts to define the catalytically active structures may be summarized in Table 10 (taken from Ref. 182), reporting the known crystallographic data in the region of the experimentally determined catalytic activity. Kolchin *et al.* [189] have determined the specific rate constants for the formation of the major products in the oxidation of propylene over α-, β- and γ-phases of bismuth molybdate, and their findings are summarized below

Phase Comp.	k_{sp}.† ($\times 10^{-2}$) min^{-1} m^{-2}		Selectivity (%)
	Acrolein	CO_2	
$Bi_2O_3 \cdot 3MoO_3$ (α)	5·29	1·85	74
$Bi_2O_3 \cdot 2MoO_3$ (β)	9·4	1·38	87
$Bi_2O_3 \cdot MoO_3$ (γ)	0·55	0·5	52

† 475°C; $t_c = 0.8_s$

Table 10. Crystallographic data of catalytically active Bimolybdate phases. (Callahan, J. L., Grasselli, R. K., Milberger, E. C., and Strecker, H. A. Ref. 182)

Phase composition	Crystal syst.	Lattice parameters (Å)	Density, g/cm³		Z	Space group	Lit. cited
			Calcd.	Pyc.			
$Bi_2O_3 \cdot 3MoO_3$ (α)	Monoclinic	a = 7·89, b = 11·70, c = 12·24, β = 116°12'				Scheelite like	[727]
	Monoclinic	a = 7·85, b = 11·70, c = 12·25, β = 116°20'	5·86	5·99	4	$C_{24}^5 = P2_{1/c}$	[503]
	Monoclinic	a = 7·719, b = 11·516, c = 11·985, β = 115°25'	6·197	6·14	4		[409], [207], [481], [35]
$Bi_2O_3 \cdot 2MoO_3$ (β)	Tetragonal	a = 11·80, c = 5·40	6·61	6·39	4	$D_{2d}^9 = I\bar{4}m2$	[503]
	Orthorhombic	a = 10·79, b = 11·89, c = 11·86			8	Pnmm	[504], [481], [35]
$Bi_2O_3 \cdot MoO_3$ (γ)	Orthorhombic (Mineral koechlinite)	a = 5·50, b = 16·16, c = 5·49			4	Cmca	[725]
	Orthorhombic	a = 5·502, b = 16·218, c = 5·483	8·281	8·26	4		[409]
	Orthorhombic γ (low temp.)	a = 5·50, b = 16·215, c = 5·485	8·25	8·0	4		[726]
	Orthorhombic γ" (meta st.)	a = 11·90, b = 11·90, (c = 5·45)	7·85	7·8	6		[726]
	Orthorhombic γ (high temp.)	a = 15·91, b = 15·80, c = 17·19	7·45	7·3	32		[726]
	Tetragonal (high temp.)	a = 3·95, c = 17·21				Isomorphous with La_2MoO_6	[189], [207], [481], [35]

The α- and β-phases, it appeared, have appreciable activity and selectivity for the conversion of propylene to acrolein, but the γ-phase displays low activity and selectivity. These results are generally similar to those of Beres et al. [204] who also found that the β-phase displayed superior properties. Batist et al. [205] whilst agreeing as to the excellence of β-bismuth molybdate, also obtained results indicating that the γ-phase in the low-temperature koechlinite modification was equally good, and in a recent paper [206], ascribed the effect of promoters for the moderately active $Bi_2O_3 \cdot 3MoO_3$ to the formation of some koechlinite. This apparent confusion may be resolved if the thermal history of the tested γ-phase is examined, since low-temperature koechlinite is transformed into a high-temperature modification (with inferior catalytic activity) at a temperature of $678 \pm 12°C$ [205], and activity will be determined by the temperature of calcination.

Significant contributions to our understanding of the excellent catalytic properties of bismuth molybdate have been made by Schuit and co-workers [205, 207, 208, 209]. Earlier work [207, 208] had indicated no connection between the presence of a particular compound and the catalytic activity, and this initiated a search for an explanation based upon the structural chemistry of the catalysts. In 1956, Zemann [725] showed that the mineral, koechlinite, has a layer structure made up of $(Bi_2O_2)_n^{2+}$ and $(MoO_2)_n^{2-}$ units connected by O^{2-} ions in the arrangement

$$(Bi_2O_2)_n^{2+} O_n^{2-} (MoO_2)_n^{2+} O_n^{2-}$$

The interspaced O^{2-} layers are arranged in such a way that in the MoO_2^{2+} planes the molybdenum ions are in octahedra sharing corners. Schuit et al. [205] subsequently suggested that all compositions of bismuth molybdate could be derived in principle from similar structural elements, Thus, if the $(Bi_2O_2)_n^{2+}$ layer is named the B-layer, and the $(MoO_2)_n^{2+}$ and O_n^{2-} layers are called the A- and O-layers respectively, then the structures of various catalyst compositions may be schematically represented.

$$2Bi_2O_3 \cdot MoO_3 \quad (Bi/Mo = 4/1) = BO\ BO\ AO\ BO\ BO\ AO \ldots$$

$$Bi_2O_3 \cdot MoO_3 \quad (Bi/Mo = 2/1) = BO\ AO\ BO\ AO \ldots$$

$$Bi_2O_3 \cdot 2MoO_3 \quad (Bi/Mo = 1/1) = BO\ AO\ AO\ BO\ AO\ AO \ldots$$

It was further suggested that the presence of catalytic activity for the selective oxidation of alkenes is connected with the presence of corner-sharing oxomolybdenum octahedra, and is largely absent in compounds containing edge-shared octahedra, a hypothesis supported by earlier work [209]. If the A- and B- layers are considered, it can be seen that, at the edges, the composition of the B-layer will be $(Bi_2O_2)_{n/2}^{2+} 2O_{2n}^{2-}$ and that of the A-layer, $(MoO_2)_{n/2}^{2+} 2O_{2n}^{2-}$. Thus, unless edge-$O^{2-}$ ions or O^{2-} ions in the terminal layer are omitted, the crystal will possess an excess negative charge. In compounds with Bi/Mo ratios of, say

1/1 or 2/1, the terminal oxygen layer actually consists of an assembly of oxo-molybdenum octahedra, so that the O^{2-} ions are at the edges of the A-layers. The omission of some O^{2-} ions, i.e. the introduction of anion vacancies connected with molybdenum cations, will lead to some of the boundary molybdenum ions being present in a tetragonally pyramidal configuration, and this, it was postulated, was the active site. The limit of catalytic activity on the bismuth-rich side is then given by the absence of anion vacancies connected to molybdenum, and hence of tetragonally pyramidal Mo–O polyhedra. Anion vacancies would still exist, but they would now be connected with Bi^{3+} ions and hence the catalyst would show the type of oxidation associated with Bi_2O_3. Unfortunately, no proof could be offered to support this attractive theory.

Recently, Mitchell and Trifiro [210] have provided some evidence in favour of this idea. A spectroscopic investigation (i.r. and u.v. reflectance) was carried out with supported and unsupported bismuth molybdate catalysts in order to identify the oxomolybdenum species present at the surface. The results obtained showed that both octahedral and tetrahedral oxomolybdenum(VI) species were present at the surface, the ratio of the tetrahedral to the octahedral form being greatest in the most active catalyst. It was also noticed that the infrared spectrum of bismuth molybdate catalysts exhibited·two strong absorptions, one at ca. 730 cm^{-1} which was assigned to molybdenum-oxygen-bridged structures and another, at ca. 850 cm^{-1}, assigned to molybdenum-terminal oxygen stretching vibrations and said to correspond to a molybdenum trioxide structure having three terminal oxygens per molybdenum atom and designated $Mo(O_t)_3$. It was the presence of this latter absorption band which was the outstanding feature of the spectra of the very active catalysts, the $Mo(O_t)_3$ species showing its maximum concentration in the most active catalyst and suggesting that this is the catalytically active species. Now $Mo(O_t)_3$ species can be formed from MoO_6 octahedra by corner-sharing, the terminal oxygen then being at the edges of the boundary layers of the MoO_6 octahedra. This structure differs from that of MoO_3 in which the octahedra share edges. Thus, although the $Mo(O_t)_3$ structure occurs in the MoO_3 structure, there are very few sites containing the structure and hence the catalytic activity of molybdenum trioxide in the selective oxidation is low. Some support for this idea can be found in the results of Matsuura and Schuit [211], who, in a recent study of the adsorption equilibria of butenes, buta-1,3-diene and water over bismuth molybdate, showed the presence of two types of site (referred to as A-centres and B-centres) on the surface. At A-centres, both desorption of water formed by hydrogen from the substrate and oxygen from the catalyst, and catalyst re-oxidation, occurred. Combustion of the hydrocarbons to carbon dioxide and water also occurred at A-centres. At B-centres, the butenes and butadiene underwent a weak, dual-site adsorption on the surface oxide ions. Mitchell and Trifiro [210] even suggested that the B-centres correspond to $Mo(O_t)_3$ species and A-centres to a bridging oxygen between either two molybdenum atoms or a bismuth and a molybdenum atom.

Another aspect of catalysis over bismuth molybdate which has aroused considerable interest is the fact that alkene oxidation can be selective both in the presence and in the absence of gaseous oxygen [185, 209]. Schuit *et al.* [209] suggested that the catalyst oxide ions were responsible for oxidation and the function of the gas-phase oxygen was merely to replenish anion vacancies created by reduction of the catalyst surface. Evidence supporting this idea has come from various sources. For example, Ashmore and his co-workers [212] have measured the electrical conductivity of pressed discs of bismuth molybdate and molybdenum(VI) oxide exposed to propylene, but-1-ene and hydrogen at 400–550°C. It was found that there was a rapid and reproducible increase in the conductivity, and a decrease to the original value on exposure to gaseous oxygen alone. In the presence of equimolar propylene/oxygen mixtures, bismuth molybdate displayed an almost constant conductivity, the value of which corresponded to a 10% reduction of the catalyst, and which varied little as the proportions of propylene and oxygen were changed. Further evidence was also obtained by Ashmore *et al.* [213] during an e.s.r. study of bismuth molybdate catalysts. When such catalysts were exposed to pure propylene, e.s.r. signals due to Mo^{5+} were detected, no such signals being found in the presence of propylene/oxygen mixtures. Similar results have also been obtained by Sancier *et al.* [214] and Burlamacchi *et al.* [215].

Perhaps the most compelling evidence for the participation of lattice oxygen in the reaction comes from studies of propylene oxidation in which isotopic oxygen has been present [198, 216, 217]. Keulks [216], for instance, examined the reactivity of the lattice oxygen in bismuth molybdate towards oxygen-exchange by circulating $^{18}O_2$ over the catalyst for a considerable time at moderate temperatures (250–500°C). No change in the gas-phase composition of oxygen was detected and, further, a circulating mixture of $^{16}O_2 + {}^{18}O_2$ did not undergo the reaction

$$^{16}O_2 + {}^{18}O_2 \rightleftharpoons 2{}^{16}O{}^{18}O$$

implying that the extent of chemisorption of oxygen on bismuth molybdate must be small, if indeed it occurs at all. Wragg *et al.* [198] have similarly found a high value (56 kcal. mole^{-1}) for the activation energy of the exchange reaction between gaseous $^{18}O_2$ and ^{16}O-labelled bismuth molybdate. Keulks also found that during the oxidation of propylene in the presence of $^{18}O_2$, only 2–2·5% of the oxygen atoms in the acrolein and carbon dioxide produced were isotopically labelled. The lack of any extensive incorporation of ^{18}O into the reaction products suggested that, not only the oxide ions in the immediate surface of the catalyst but those from many (up to 500) sub-surface layers participate. Thus the diffusion of oxygen from the surface and into the bulk of the catalyst and from the bulk back to the surface, must be rapid, an idea first put forward by Schuit [218].

In summary, therefore, it appears that the selective oxidation of propylene to acrolein initially involves the irreversible dissociative chemisorption of propylene, according to the equation

$$C_3H_6 + Mo^{6+} + O^{2-} \rightarrow (C_3H_5 \ldots Mo)^{6+} + OH^- + e^-$$

Thus Krivanek et al. [219] have obtained a value for the heat of adsorption of between 22·3 and 11·4 kcal. mole^{-1}. In the light of later results, Krivanek and Jiru [220] have revised the upper value to 19 kcal. mole^{-1}. The site of adsorption may be a molybdenum atom having three terminal oxygen atoms attached to it [210]. According to Peacock et al. [213], the bond between the molybdenum ion and the allyl species is likely to involve $\sigma\pi$-bonding, the four-centred σ-bonding molecular orbital being formed from the filled π-bonding orbital of the allylic species and the initially empty $4d_{z^2}$ atomic orbital of the molybdenum cation. The three-centred π-bonding molecular orbital is formed from the empty d_{xz} atomic orbital of the molybdenum and the non-bonding orbital of the allyl ligand. Of the two molecular orbitals, the σ orbital is filled but the π ligand-cation bonding orbital contains only one electron. Such bonding would result in a partial positive charge on the chemisorbed allyl species and also allow for the reduction of an adjacent Mo^{6+} cation to Mo^{5+} by the freed electron. It was also thought that, although further oxidation of the allyl species or chemi-sorption of propylene on Mo^{5+} may lead to a further reduction to Mo^{4+}, it is more likely to involve a bismuth cation; and it appeared that reduction of Bi^{3+} could be associated with the further oxidation of the allyl species by coupled reactions leading to net processes such as

$$(C_3H_5 \ldots Mo)^{6+} + OH^- + O^{2-} + 3Bi^{3+} \rightarrow C_3H_4O + Mo^{6+} + H_2O + 3Bi^{2+}$$

Now, the electron produced in the initial step

$$C_3H_6 + Mo^{6+} + O^{2-} \rightarrow (C_3H_5 \ldots Mo)^{6+} + e^- + OH^-$$

may be localized on, and reduce a second Mo^{6+}, or it may be localized on the chemisorption complex. In the latter case, the allyl species would be more strongly bonded to the molybdenum, but the C—C bonds would probably be weakened and C—C fission within the allyl species would lead to an increased rate of production of oxides of carbon by a route not involving acrolein oxida-tion. However, if the electron is localized on bismuth ions, the allyl species will remain weakly bonded to molybdenum and fission will be less likely; thus the presence of bismuth allows a greater fraction of molybdenum ions to remain in, or return to a higher oxidation state than in pure MoO_3. This stability towards reduction has also been suggested [214] to be critical in bismuth molybdate catalysts with the best activity and selectivity.

(e) *Miscellaneous investigations.* Very recently, the results have been reported of three investigations of the influence on subsequent catalytic activity, of the experimental conditions (pH, temperature of precipitation, etc.) during the preparation of Bi–Mo–O catalysts [221–223]. A similar investigation had been made earlier by Beres *et al.* [204]. Particularly important, it appears, are the conditions in the "wet stage" of the preparation, and the slurry reactions between the precipitated oxyhydrates. Thus Schuit and co-workers [221] found that active catalysts of composition Bi/Mo = 2/1, have to be prepared by a slurry reaction starting essentially from the compound $(BiO)(NO_3)$:

$$2(BiO)(NO_3) + H_2MoO_4 \rightarrow (BiO)_2MoO_4 + 2HNO_3$$

and have summarized the connections between the various forms of catalyst. This may be summarized below (scheme after Schuit *et al.* Ref. 221).

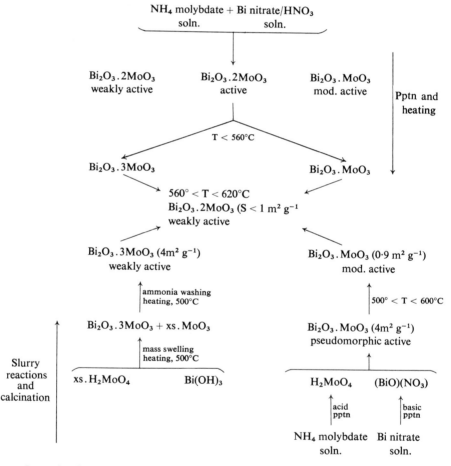

Investigations have also been made of the effect of the texture of BiMoP catalysts on the yield of oxygen-containing compounds but these studies have been rather inconclusive. Thus Azimova *et al.* [225] found that catalysts having high porosity (68 %) and high surface area (190 m^2 g^{-1}) were most active and selective, whereas Lemberanskii and co-workers [226] had previously reported the excellence of low porosity (21 %), low surface area (0·6–1·2 m^2 g^{-1}) catalysts. Recently, Gagarin *et al.* [445] have obtained a quantum chemical description of the behaviour of propylene during oxidation over bismuth molybdate catalysts. Calculations tend to confirm the generally accepted view that there action proceeds via an allylic intermediate.

Finally, Paryjczak *et al.* [227a, b] have determined the influence of preparative conditions (calcination temperature etc.) on the specific surface areas of Bi_2O_3–MoO_3 and Bi_2O_3–MoO_3–P_2O_5 mixtures.

C. SnO_2–MoO_3 Catalysts

Tin(IV) oxide is not a particularly good catalyst for the selective oxidation of propylene. For example, Tan *et al.* [228] found that, at temperatures below 300°C, less than 5 % of propylene was converted to products which were exclusively carbon monoxide and dioxide; at temperatures above 300°C, the degree of propylene conversion increased, but the products were still mainly the oxides of carbon, although small amounts of acetone, acrolein, acetaldehyde and acetic acid were also formed. Thus at 315°C, the selectivity of propylene oxidation towards acrolein was only 4·7 %, and fell to 1·1 % at 415°C. However, Buiten [229] found that in the presence of steam (propylene : air : steam = 1 : 10 : 5), propylene was oxidized at temperatures between 297 and 354°C to acrolein with a selectivity of between 40 and 45 %. Similarly Lazukin *et al.* [203] found that, at 450°C, propylene was oxidized to acrolein with a selectivity of about 26 %, although it is not clear whether water vapour was present in the feed-gases.

Molybdenum(VI) oxide, although much less active than tin(IV) oxide, appears to be much more selective in its behaviour. For instance, Lazukin [203] found that, although its activity was fourteen times less than tin(IV) oxide, molybdenum(VI) oxide could catalyse the air-oxidation of propylene with a selectivity towards acrolein of 87 %. Buiten [229], on the other hand, has found the oxidation to be much less selective (30–50 %) towards acrolein at 497°C, a range of other products including acetaldehyde, acetone, acrylic acid, acetic and maleic acid, also being formed.

In combination, however, tin(IV) oxide and molybdenum(VI) oxide appear to produce a peculiar and apparently very versatile catalyst which has been the subject of a number of investigations. To illustrate the versatility, examples may be taken from the work of Tan *et al.* [228], Buiten [229] and Lazukin and co-workers [230]. Tan *et al.* found that a mixture of tin(IV) oxide and molybdenum(VI) oxide (Sn : Mo = 9 : 1) would, in the presence of water vapour and at fairly low temperatures (100–160°C), convert propylene to acetone with a selectivity of almost 90 %. (This oxidation, however, will be thoroughly discussed in Section 3.5.) Lazukin *et al.* [230] reported the selective (79 %) oxida-tion of propylene to acrolein over catalysts with ratios of 9:1 and 1:9. Oddly, however, Pluta and Blasiak [231] have also examined the effectiveness of molybdenum(VI) oxide–tin(IV) oxide catalysts for the oxidation of propylene to acrolein and found that, in the temperature range 310–420°C, the most effective compositions ($83 \cdot 2$ $MoO_3/17 \cdot 7$ SnO_2 and $63 \cdot 5$ $MoO_3/36 \cdot 5$ SnO_2) were, even so, not particularly good, a mere 8 % of the propylene being converted with a selectivity towards acrolein of only 40–50 %. (The addition of $2 \cdot 66$ mole % B_2O_3 to the feed gases, however, somewhat improved the activity and selec-tivity.)

The tin(IV) oxide–molybdenum(VI) oxide system has been studied in some detail. Thus, Gerei and co-workers [232] investigated the system by means of i.r. spectroscopy (600–1200 cm^{-1}) and found that, depending on the method of preparation, oxomolybdenum–oxotin compounds were formed. Catalysts were prepared by three methods: (i) by mechanically mixing tin(IV) oxide and molybdenum(VI) oxide (group A); (ii) by co-precipitation from mixed tin(IV) chloride and ammonium molybdate solutions (group B); and (iii) by mixing tin(IV) hydroxide and molybdic acid (group C). Catalysts prepared by all three methods were calcined in air at 600°C and their spectra were examined. It was found that group A mixtures were only mechanical mixtures of the oxides whilst groups B and C showed absorptions indicating compound formation. In the case of group B catalysts, compounds with the suggested structure

$$O_nMo-OSn(O)-O-MoO_n \qquad (n = 1 \text{ or } 2)$$

were formed, whilst group C catalysts contained compounds of the suggested structure

$$O_nMo \underset{O}{\overset{O}{\diagup \diagdown}} Sn \underset{O}{\overset{O}{\diagup \diagdown}} MoO_n$$

Similarly, Lazukin *et al.* [230] found that catalysts of Sn:Mo ratio 50:50, 35:65, 25:75 and 10:90, heated to 600°C, consisted of a mixture of molybden-um(VI) oxide, the solid solution molybdenum(VI) oxide in tin(IV) oxide with a Sn:Mo ratio of 65:35, and a compound of formula $SnO_2 \cdot 2MoO_3$. Neither the latter compound nor any other compounds were detected in Buiten's study [229], although it was discovered that the peculiar catalytic activity of tin–molybdenum oxide catalysts in the presence of steam at about 450°C was due, not to the formation of a tin–molybdenum compound in the *bulk*, but to the coverage (effectively a monolayer) of the tin(IV) oxide surface with a molyb-denum–oxygen compound, $MoO_2(OH)_2$.

Recently, Firsova *et al.* [233] have reported the results of a study, using Mössbauer spectroscopy, of the effects of adsorbing propylene and acrolein on the surface of a solid solution of tin(IV) oxide in molybdenum(VI) oxide (Sn:Mo = 2:1). Prior to adsorption it was found that the γ-resonance spectrum of the Sn–Mo compound was very similar to that of pure tin(IV) oxide, but, after a preliminary treatment of the catalyst with oxygen, adsorption of propylene produced a change in the spectrum corresponding to the reduction of Sn(IV) to Sn(II). Further experiment revealed that propylene and acrolein are adsorbed on the surface of Sn–Mo catalysts by Sn—O bonds, giving a complex of the type

$$C_2H_5-C\overset{O}{\underset{O}{\diagup}}\overset{Sn}{\underset{O}{\diagdown}}\overset{O}{\underset{O}{\diagup}}C-C_2H_5$$

The valence change of the tin ion was observed only in Sn–Mo catalysts, and not with pure tin(IV) oxide, indicating the great importance of molybdenum during complex formation. A possible explanation for this is that molybdenum(VI), unlike Sn(IV), can form π- or π-allyl-complexes with olefins, undergoing reduction to either Mo^{5+} or Mo^{4+} and electrons may then be transferred to Sn^{4+} ions either through the solid lattice or inside the complex sphere. Margolis [196] has indicated a scheme for propylene oxidation over tin-molybdenum catalysts, based upon this idea:

Step 1. $Mo^{VI} + C_3H_6 + O^{2-} \longrightarrow Mo^V \dots C_3H_5 + OH^- \dots (Mo^V)$

Step 2. $Mo^V(OH^-) + O_2(ads.) + Mo^V \dots C_3H_5 \longrightarrow Mo^V(OH) \dots O_2 \dots Mo^V \dots C_3H_5$

$$(OH^-)Mo^V\overset{}{\underset{C_3H_4.Mo^V}{\diagup}}O_2 \longleftarrow (OH^-)Mo^V\overset{}{\underset{C_3H_5.Mo^V}{\diagup}}O_2 + O^{2-}$$

Step 3.

$$Sn^{4+}:O_2\overset{Mo^V(OH^-)}{\underset{Mo^V \dots C_3H_4}{\diagup}} \longrightarrow Sn^{II} \dots O_2\overset{Mo^V(OH^-)}{\underset{Mo^V \dots C_3H_4}{\diagup}} \longrightarrow \boxed{Acrolein}$$
(Rate-deter. step)

Step. 4. $Sn^{II}O \xrightarrow{+O_2} Sn^{IV}O_2$

Step 5. $(OH^-) + (H^+) \longrightarrow H_2O$

Basically, the first step in the scheme is the adsorption of propylene on Mo(VI) ions to form a π-allyl complex and a hydroxyl group bound to a Mo(V) ion. The second hydrogen atom is then removed from the allyl intermediate by reaction with oxygen. The oxygen-hydrocarbon complex bound to Mo(V) then reacts with a tin(IV) ion which is then reduced [189]. Decomposition of the tin-molybdenum-hydrocarbon complex then results in the adsorption of acrolein and the eventual restoration of the initial charges on the tin and molybdenum ions.

Finally, Orlov *et al.* [234] have studied the adsorption of propylene on tin–molybdenum catalysts by means of i.r. spectroscopy. Propylene was adsorbed on degassed catalysts at 20°C and at 20 and 200°C on a surface pre-treated with oxygen. In the absence of oxygen gas, bands at 2950 and 2870 cm^{-1} (symmetrical and asymmetrical C—H vibrations in —CH$_3$), 2925 cm^{-1} (asymmetric C—H vibrations in —CH$_2$—) and 1456 and 1430 cm^{-1} (deformation vibrations of C—H in —CH$_2$— and —CH$_3$, respectively), were observed. On the oxygen-treated surface, at 20°C, bands at 3000 cm^{-1} (valence vibrations of CH in either alkenes or cyclanes), 2930 cm^{-1} (asymmetric vibrations of —CH$_2$—) and 1456 cm^{-1}, were observed; at 200°C, —CH— bands were not observed and absorption at 1507 cm^{-1} (due to C=C vibrations with a coordinate bond to the surface) and 1690 cm^{-1} (carbonyl vibrations), appeared.

D. *Tin Oxide–Antimony Oxide Catalysts*

Commercially, tin–antimony oxide catalysts have been developed by Hadley and co-workers [235], not only for the oxidation of propylene to acrolein, but also for the ammoxidation of propylene to acrylonitrile and the oxidative dehydrogenation of butenes to buta-1,3-diene. However only the first process will be considered in this section.

The activity of pure antimony oxides in the selective oxidation of propylene has been studied by Lazukin *et al.* [203] and Belousov and Gershingorina [236]. Lazukin reported that antimony(IV) oxide was fairly inactive (activity = 0·13, cf. 21·64 for Bi/Mo and 0·46 for MoO$_3$) but fairly selective (85% cf. 94% for Bi/Mo and 87% for MoO$_3$). Belousov and Gershingorina [236] have also examined the oxidation reactions of propylene over antimony oxides containing the cation in various valence states. Generally, the oxidation products were carbon dioxide, water, aldehydes (acrolein, acetaldehyde and propionaldehyde) and acids (acetic and acrylic), but the product spectrum and selectivity depended strongly upon the valence of the cation. Thus, acids were formed only with antimony(V) and antimony(III) oxides, propionaldehyde only with antimony(V) oxide, and, although acrolein was formed in the presence of all oxides, it was observed with antimony(IV) oxide only at high temperatures. It was also found that at 300°C, antimony(V) oxide was 15 times more active than antimony(IV) oxide and some 50 times more active than antimony(III) oxide; the selectivity of antimony(V) oxide towards aldehydes (70%) was also greater than that of antimony(IV) oxide (54%) or antimony(III) oxide (10%). However, selectivity towards acids was greater (35%) for antimony(III) than for antimony(V) oxide (7%). Tin(II) and tin(IV) oxides were also examined and it was found that both oxides, like the antimony oxides, adsorbed propylene strongly on the surface in a catalytically inactive form. On tin(II) oxide, the oxidation of propylene yielded mainly acids (selectivity 40–80%) and no aldehydes, whereas tin(IV) oxide gave aldehydes (acetaldehyde and acrolein, ratio = 1:2) and acetic acid. The selectivity of tin(IV) oxide towards the aldehydes was strongly dependent on temperature, the maximum selectivity (*ca* 50%) being found at 280°C, and thereafter falling to a fairly constant value of *ca*. 10% at 330°C and above. Activation energies for propylene oxidation over the various oxides were as follows:

Oxide	E_A *(kcal. mole^{-1})*
Sb_2O_5	9–10
Sb_2O_4	11–12
Sb_2O_3	14–16
SnO_2	25–27
SnO	10

Lazukin and co-workers [203, 237] have also prepared mixed tin–antimony oxides (Sn:Sb = 9:1 to 1:9) and examined the activity and selectivity towards oxidation. It was found [203] that at 450°C the mixed oxides (Sn:Sb = 9:1) were slightly more active (1·41) but considerably more selective than pure Sb_2O_4. In the later study [237], catalysts (prepared by the air calcination at 1050°C of mixed tin(II) and antimony(III) hydroxides) were examined by a range of physical techniques. With a Sn:Sb ratio of 75:25, a solid solution was formed. All catalysts had *n*-type semiconductivity and contained antimony cations in the III- or IV-valent state. Over these catalysts, the air oxidation of propylene was selective (95% towards acrolein and acetaldehyde) and, so long as the solid solution was present, the composition of the catalyst had little effect on overall selectivity.

Godin *et al.* [238] have also reported results of studies of the kinetics and mechanism of propylene oxidation over tin–antimony oxide catalysts. Catalysts were prepared by mixing the oxides in the required proportions and calcining in air at 850°C. X-ray analysis revealed the presence in the catalysts, of an inhomogeneous mixture of tin(IV) oxide, α-Sb_2O_4 and small amounts of β-Sb_2O_4, but there was no indication of compound formation. In accordance with the results of Lazukin [237], neither catalyst composition nor temperature had any marked effect on the selectivity of the reaction towards acrolein (Tables 11 and 12). It was also found that the addition of only 6·8% Sb to tin(IV) oxide increased the conductivity by three orders of magnitude to give a semiconductor with a very low temperature coefficient and this was attributed to the substitution of Sb^{5+} for Sn^{4+} ions in the tin(IV) oxide lattice, creating firstly Sn^{3+} ions, which acted as electron donors and then giving free electrons and a rise in conductivity:

$$Sn^{4+}_{1-2x}Sb^{5+}_x Sn^{3+}_x O^{2-}_2 \rightleftharpoons Sn^{4+}_{1-x}Sb^{5+}_x O^{2-}_2 + xe^-$$

Table 11. The variation of selectivity towards acrolein with temperature and composition. (Fixed C_3H_6 conversion ($=20\%$) but variable contact time [237])

Catalyst composn. (atom % Sb)	0	6·8	12·4	50·0	78·6	93·7	100
Reaction temp. (°C)	381	336	339	339	282	478	520
Selectivity (%)	11	63	53	59	72	76	5

Table 12. The variation of selectivity towards acrolein with composition (Fixed temp. (330°C), propylene conversion (16%) but variable contact time)

Catalyst composn. (atom % Sb)	0	6·8	12·4	50·0	78·6	93·7	100
Selectivity (%)	10	59	58	60	71	75	—

Like the oxidation of propylene over bismuth molybdate, the oxidation over tin–antimony oxides is first-order with respect to propylene, but there is also a dependence on oxygen. This has been confirmed recently by Belousov and Gershingorina [239]. The possible participation of lattice oxygen in the oxidation of propylene over antimony(V) and antimony(III) oxide/tin(IV) oxide catalysts was examined but no support was found for the redox mechanism observed with bismuth molybdate.

Godin *et al.* [238] also derived values for the activation energy for propylene oxidation over various catalyst compositions. Their values are given below

Catalyst composn. (atom % Sb)	E_A (kcal. mole^{-1})
6·8	12·1
21·4	16·3
50·0	17·1
78·6	13·1
93·7	13·0

Studies using 1-^{13}C-labelled prop-1-ene have also confirmed an oxidation mechanism involving a π-allyl species and a mechanism has been proposed [238] similar to that of Batist *et al.* [209] for bismuth molybdate catalysts:

$$C_3H_6 + A.V. + O^{2-} \rightarrow C_3H_5^- + OH^-$$
$$Sb^{5+} + C_3H_5^- \rightarrow C_3H_5...Sb^{4+} \tag{1}$$

$$C_3H_5...Sb^{4+} + 2O^{2-} \rightarrow C_3H_4O + Sb^{3+} + OH^- + 2e^- + 2A.V. \tag{2}$$

$$2OH^- \rightarrow O^{2-} + A.V. + H_2O \tag{3}$$

$$\tfrac{1}{2}O_2 + Sb^{3+} + A.V. \rightarrow O^{2-} + Sb^{5+} \tag{4}$$

$$\tfrac{1}{2}O_2 + 2e^- + A.V. \rightarrow O^{2-} \tag{5}$$

A.V. = anion vacancy

Results have also been reported [240] of experiments on the effects of the chemisorption of propylene and oxygen on tin(IV) oxide and tin(IV)–antimony-(IV) oxides, with particular reference to changes in electron work function and electrical conductivity. At 200°C and on both surfaces, propylene and oxygen were rapidly adsorbed. In the case of tin(IV) oxide, adsorption of propylene upon irreversibly pre-adsorbed oxygen yielded a surface complex having a negative change and containing an excess of oxygen with respect to carbon. However, in the case of tin-antimony oxide mixtures, the surface complex had a positive change and contained an excess of carbon relative to oxygen. The latter type of complex is similar to that formed on the surface of other selective oxidation catalysts such as $Bi_2O_3 \cdot 3MoO_3$ and $SnO_2 \cdot MoO_3$. Margolis et al. [241] also employed Mössbauer spectroscopy to study the surface complexes formed during the adsorption, at 200°C, of propylene, acrolein and their mixtures with O_2, on Sb:Sn (2:1) and $FeSbO_4$ catalysts. Spectra were obtained at 77 and 300K. Stable surface compounds were observed, and it was concluded that the chemisorption of both acrolein and propylene takes place initially on antimony atoms, forming a system containing equal amounts of Sb^{5+} and Sb^{3+} ions. Apparently, the role of both Sn^{4+} and Fe^{3+} ions is to shift the equilibrium in the initial systems towards Sb^{5+} and to stabilize the Sb^{5+}/Sb^{3+} ratio during catalysis.

In other studies, Wakabayashi et al. [242] oxidized propylene at 460°C over mixed Sn/Sb oxides supported on alumina. The optimum conversion of propylene to acrolein (>70%) occurred over a catalyst with an atomic ratio of 3Sn:1Sb which had been presintered for 3 hours at 1000°C. The sintering temperature for the catalysts strongly influenced the yield of acrolein. X-ray studies suggested that the catalysts consisted of solid solutions of antimony oxide in tin oxide. Further work by Wakabayashi et al. [243] on the physico-chemical properties of the catalysts revealed that, when antimony oxide was added to tin(IV) oxide, the electrical conductivity increased rapidly to a maximum at 3 atom % Sb, declining slightly thereafter. The surface area also rapidly increased to a maximum at 3 atom % Sb. The activation energy for acrolein formation remained virtually constant (ca. 17 kcal. mole^{-1}) over the whole range of compositions, although the apparent activation energy for carbon dioxide formation decreased initially as the antimony content of the catalysts was increased before reaching a constant value. The pure oxides and oxide mixtures were also reduced in hydrogen and it was found that the activation energy for the reduction of pure tin(IV) oxide was 28 kcal. mole^{-1}, for pure antimony oxide it was 30 kcal. mole^{-1} and for tin oxide containing 25 atom % Sb it was 14 kcal. mole^{-1}. Finally, a study has been made by Roginskaya et al. [244] of mixed Sb_2O_3–SnO_2 catalysts (Sn:Sb atom ratios = 0:1, 1:8, 1:4, 1:2, 1:1, 5:1 and 1:0) calcined at 500, 700 and 900°C. Using X-ray and i.r.-methods it was found that when the tin content of the catalysts was increased, the phase composition changed from the defective Sb_2O_4 structure, over a multiphase structure containing a solid solution of Sb_2O_4 in SnO_2 to the tin(IV) oxide structure. The catalysts were screened for their activity in the oxidative dehydrogenation of butenes and in the ammoxidation of propylene and Roginskaya concluded that the catalytic activity was due to the presence of antimony oxides containing low-valence antimony ions.

In conclusion, it is interesting to note that Dewing et al. [245] claim that antimony–tin oxide catalysts (46% Sb + 25% Sn) can be used at 490°C for the direct oxidation of propane to acrolein, although the yield is fairly low (ca. 30%).

E. Antimony Oxide–Uranium Oxide Catalysts

Catalysts of this type have been developed by Sohio [246]. Although the combination is probably active and selective for the oxidation of propylene to acrolein, little work on this particular conversion has been reported in the literature. Results have been obtained using this catalyst in the oxidative dehydrogenation of but-1-ene [247], and iso-pentene isomers [248] and in the ammoxidation of propylene [249, 250, 251] but these will be dealt with in appropriate sections.

F. Bismuth Oxide–Antimony Oxide Catalysts

The system has been examined by Lazukin and his co-workers [252] and Ohdan et al. [253]. Lazukin et al. [252] examined the phase composition and activity of such catalysts towards propylene oxidation at temperatures between 380 and 480°C. The activity of the mixed oxide catalysts lay between that of the components and, although catalysts with Bi:Sb > 1 catalysed the extensive oxidation of propylene, oxidation over catalysts with higher antimony content was selective towards acrolein. Even on the most active catalysts (Bi:Sb = 1:3), the rate of acrolein formation was low (0.10–0.30×10^{-7} mole propylene m^{-2} s^{-1} cf. 21.64×10^{-7} mole m^{-2} s^{-1} for Bi:Mo = 1:1). Lazukin et al. [252] reported that the catalysts of high bismuth content (up to 2Bi:1Sb) are solid solutions of antimony (IV) oxide in Bi_2O_3. Catalysts of composition Bi:Sb = 1:1 are composed of bismuth antimonate $BiSbO_4$, whilst catalysts of composition 50:50 > Sb:Bi > 28:72 consist of two solid phases—Sb_2O_4 and a saturated solid solution of Sb_2O_4 in $BiSbO_4$. It was concluded that Sb_2O_4 and its solid solutions with $BiSbO_4$ were the active components of the mixed oxide catalysts. Similarly, Ohdan et al. [253] found that, if mixed bismuth and antimony oxides were calcined at temperatures in excess of 470°C, two compounds were formed, namely $Bi_2O_3 \cdot Sb_2O_5$ ($\equiv BiSbO_4$) and $3Bi_2O_3 \cdot Sb_2O_5$ ($\equiv (BiO)_3SbO_4$).

Ohdan and co-workers [254, 255] have also examined the Bi–Mo–Sb oxide system and found that the system $BiSbO_4$–Bi_2O_3–$3MoO_3$ was an even more active and selective catalyst for the oxidation and ammoxidation of propylene than the binary Bi–Mo system. Thus, in a recent patent, Yamada and co-workers [256] claim that catalysts containing 5–60% Mo, 6–70% Sb and 25–75% Bi are, at 150°C, effective for the ammoxidation of propylene, 83.6% of the propylene being converted to products containing 91.3% acrylonitrile.

Earlier work by Ohdan [254] had shown that the Bi–Mo–Sb system, prepared by treating antimony oxides with bismuth molybdate and bismuth(III) nitrate and calcining at 540°C (16h), contained the compounds $Bi_2O_3 \cdot 3MoO_3$, $Bi_2O_3 \cdot MoO_3$, $Bi_2O_3 \cdot Sb_2O_5$ and $3Bi_2O_3 \cdot Sb_2O_5$.

G. Other Mixed Oxide Catalysts

The patent literature contains many claims for catalyst formulations effective for the conversion of propylene into acrolein. Some are excellent whereas others are less good; some recent catalysts which appear to be useful are given in Table 13.

Outside the patent literature, too, work has been reported on the oxidation of propylene over catalysts not so far covered. For example, Zhiznevskii et al. [305] studied the kinetics of propylene oxidation over Fe–Te–Mo catalysts at 420°C and obtained the following data:

Rate of acrolein formation $= k_1[C_3H_6]^{0.7}[H_2O]^{0.7}$ mol. $l^{-1} s^{-1}$

Rate of CO_2 formation $= k_2[\overset{\cdot}{C_3}H_6][O_2]^{0.15}[H_2O]^{0.1}$ mol. $l^{-1} s^{-1}$

Rate of CO formation $= k_3[C_3H_6]^{0.7}[O_2]^{0.2}[H_2O]^{0.7}$ mol. $l^{-1} S^{-1}$

where

$$k_1 = 2.01 \times 10^3 \exp(-14,388/RT)$$
$$k_2 = 3.16 \times 10^7 \exp(-25,373/RT)$$
$$k_3 = 3.64 \times 10^4 \exp(-17,398/RT)$$

Vanadium/molybdenum oxide catalysts have been examined by Wakabayashi and Kamiya [306] and by Golodets et al. [307]; bismuth/tungsten oxide catalysts by Ohdan et al. [308]; Fe–Sb–Mo oxide catalysts by Kuchmii et al. [309]; and Fe–Li–As oxide catalysts by Ishikawa et al. [310].

3.3. To Acrylic Acid

A. Processes

Acrylic acid has a use as a monomer for the production of polyacrylic acid and other acrylic polymers and is, therefore, a commercially desirable product. It has been known for some time that acrolein can be converted to acrylic acid by heterogeneous oxidation [311, 312] and recently catalysts have been described having high activity and great selectivity [313, 314] for this process. For example, a 99% conversion of acrolein with a selectivity towards acrylic acid of 85%, is claimed for a Mo–Tl–Re (1:0.1:0.1) oxide catalyst at 310°C and in the presence of steam [314]. On the basis of these observations, therefore, it is possible to convert propylene to acrylic acid, but this involves a two-stage process. A typical example of such a process is that developed by W. R. Grace and Co. [315]. A mixture of propylene, air and steam is passed through two reactors in series, the first (at 440°C) contains a bismuth molybdate catalyst which converts propylene to acrolein, whilst the second reactor (at 426°C)

Table 13. Some commercial catalyst configurations for the active and selective oxidation of propylene to acrolein.

Authors	Catalyst composition	Feed-gas composition				T(°C)	C(%)	Y(%)	S(%)	Ref.
		C_3H_6	O_2	N_2	steam					
B.A.S.F. A.-G.	Mo+W (1:2) + 1·3% TeO_2	5·4%	64·6% (air)		30%	300	84	65		[257]
Callahan and Gertisser	Sb+Fe (8:8:1) oxides/SiO_2		various			413		63		[258]
Deutsche Erdoel A.G.	Bi+Mo+P+Fe	20	106 (air)		26	350–500	15	85–90		[259]
Shell International Res.	Bi+Mo+0·1 to 5% Cu†		—		—	350–450				[260]
Engelbach, Krabetz & Buechler	Mo+W+Te (1:2:0·06)	1	2·4 (air)		5·6	330	82	67		[261]
S.N. des Pétroles d'Aquitaine	Mo+Te+P/SiO_2 (10–40:2–50:1)	15%	71% (air)		14%	415	38		84	[262]
Furuya	Bi+Mo (2:1)+Bi-sub. zeolite	1	1 (air)		—	400	45	70		[263a]
Furuya	Bi+Mo (2:1)+Mo-sub. zeolite	1	1 (air)		—	400	20	80		[263b]
Furuya	Bi+Mo (2:1)+Pd-sub. zeolite		—		—	130	15	88		[263c]
Ikeda, Ishii and Nakano	Sn+Sb+V+Fe+Bi (1:2:1:0·5:0·01)	6·0	12·0	67·0	15·0	270	62		73	[264]
Ito, Nakamura and Inoue	Cr+Mo+Te+Cd (2:14, 74·9, 3·7, 0·7) at%	4%	50% (air)		46%	374 / 383	93 / 96		83 / 76	[265]
Ito, Nakamura and Nakano	Fe+Ni+Cr+Bi+Mo+Sn +B+Ti/SiO_2		—		—	295–326			74–85	[266]
Ito, Inoue and Nakamura	Cr+Al+Mo+Te (2:0·7:7:0·35)	4%	50% air		46%	377 / 384	90 / 95		84 / 83	[267a]
Same authors	Cr+Cd+Mo+Te (2:0·7:7:0·35)	4%	50% air		46%	373 / 383	93 / 96		83 / 76	[267b]
Koch	Ca+Mo (1:1·02)+Bi	2	—		3	450	26			[268]
Mitsubishi Rayon Co. Ltd.	Mo+Te (1 of Sn, Sb, Pb, W, V, Cr, Ag, Cd, Ti, Cu, Ni, Mn, Zn, Bi)	10·7%	59·3% air		30%	400	75	67		[269]
Nakayama, Ogawa and Asao	W+Mo+Te	76	624 (air)		300	345		76		[270]
Ono and Ikumi	V+Fe+Bi+Mo (1·7–51, 0·7–32, 13–65, 7–70 at %)	10%	60% air		30%	400	55		89	[271]
Nakayama, Ogawa and Uschida	Mo+Sb+Bi (3:1:3)	7·4%	62·6% air		30%	470	39		82	[272]
Nemec and Schlaefer	Cu+Mo (24%, 49%) + CuTe (0·11%)	1	4·6 air		4	455	30	74		[273]
Nippon Kayaku Co. Ltd.	Ni+Co+Mo+O (4·5, 4, 12, 53)+Te, Bi+As	1	10 air		6	330	94		81	[274]

Author	Catalyst					Temp.				Ref.
Parks-Smith	TeO$_2$ (+ MoO$_3$)	2.5%	48.5% air		49%	390 395	50	89		[275]
Rohm und Haas G.m.b.H.	Bi + Mo + sulphate ion (20–31 pts)	15 l h^{-1}	71.5 l h^{-1} air		20 l h^{-1}	430	73		86	[276]
Rohm und Haas G.m.b.H.	MnSO$_4$ + TeO$_2$ + MoO$_3$ (0–611)	1	14.3 air		6.8	420			86	[277]
Takenaka and Yamaguchi	Ni$_{4.5}$Co$_4$FeBiAsMo$_{12}$O$_{53}$/SiO$_2$	1	10 air		6	330	94	76		[278a]
Same authors	Ni$_{4.5}$Co$_4$FeBiP$_{0.08}$Mo$_{12}$O$_{31}$/SiO$_2$	1	10 air		6	310	96		83	[278b]
Shell International Res.	Bi + Mo/ZrO$_2$(+ Cu)	91%	6.2% air		—	500	42		79	[279]
S.N. des Pétroles d'Aquitaine	MoO$_3$ + P$_2$O$_5$ + As$_2$O$_5$	14	71 air		15	400			97	[280]
Uda, Sakurai and Abe	a Fe b As c Mo d Bi e O, a = 0.5–10, b = 0.1–10, c = 12, d = 0.5–6, e = 38–81	45 cm^3 s^{-1}	255 cm^3 s^{-1} air		9.2 g h^{-1}	430	82	77		[281]
Sudo, Kita and Tamayori	P + W + Mo + Te + 2 to 50% Co, Ni, Fe, Zn, Bi or Zn + Co	8%	62% air		30%	325 350 355 370 350	97 93 96 94 64		73 70 70 68 70	[282]
Takayama and Ikeda	TeO$_2$ + W, V, As or Sn and/or Sb	10%	60% air		30%	430				[283]
Tokutaniyama, Kato and Baison	Sn + Sb + U	5%	75%	87.5%	—	430	89		81	[284]
Takayama, Nakauama and Yoshizawa	W + Te (cont. Zn, Cd or Hg) + P	8.7%	51.3% air		40%	380	81		67	[285]
Hirota, Kashiwabara and Nakamura	Mo + Bi + (Co + Ni) + Sb + Al 9.1, 2.2, 0.5, 0.5, 4.4, 4.4	1	0.47 air		—	500	24		73	[286a]
	Mo + Bi + V + Sb + Al 9.1, 3.8, 0.9, 4.12, 41.2	1	0.50 air		—	495	25		72	[286b]
	Mo + Bi + V + Al + P 10.9, 7.6, 1.0, 412, 412	2	1 air		16	500	27		60	[286c]
Takenaka, Kido, Shimabara and Ogawa	NiCo$_3$Fe$_2$BiP$_2$K$_{0.2}$Mo$_{12}$O$_{50}$/SiO$_2$	1	10 air		5	305	96		88	[287]
Yamagashi, Sakakibara and Yasube	Fe + Bi + Mo + As/SiO$_2$	50 cm^3 min^{-1}	290 cm^3 min^{-1}		10.4 cm^3 min^{-1}	395	85	70		[288]
Takenaka and Yamaguchi	Ni$_{10}$Co$_{0.3}$FeBiPMo$_{12}$O$_{51}$	2.1 l h^{-1}	16.0 l h^{-1}		10 l h^{-1}	310	95	90	75	[289]
Ohara, Ueshima et al.	Co$_4$FeBiW$_2$Mo$_{10}$Si$_{1.35}$K$_{0.06}$OX	4	51 air		45	320	97		93	[290]
Shiraishi et al.	Mo$_{12}$BiFe$_2$Mg$_{1.5}$Co$_{1.5}$Ni$_6$Tl$_{0.1}$ PO$_{52.2}$	1	7 air		7		87		98	[291]

Table 13—continued

Authors	Catalyst composition	Feed-gas composition				$T(°C)$	$C(\%)$	$Y(\%)$	$S(\%)$	Ref.
		C_3H_6	O_2	N_2	steam					
Callahan and Gertisser	$Sb_a^{II-V}Fe_b^{II-III}O_c$/SiO$_2$	1	10 air		Present	413		63	—	[292]
Deutsche Gold- u. Silber	Ni, Co, Fe, Bi, P, Mo, Sm, O/montmorillonite	1	10 (air)		2	360	90	72	—	[293]
Celanese Corp.	Mo (100):W (50–250):Te (0·3–3·5):O (400–1100)	13·4	202 (air)		122 ml/min	450	93	64	—	[294]
Ube Industries Ltd.	70:30 w% BiSbO$_4$:Bi$_2$W$_3$O$_{12}$	1	1	—	6	470	62·2	—	76·6	[295]
	70:30 w% BiSbO$_4$:Bi$_2$(MoO$_4$)$_3$	1	1	—	6	470	62·2	—	76·6	[296]
	BiSbO$_4$ + BaMo$_3$O$_{10}$	1	7·5 (air)		2	470	40·6	—	89·8	[297]
Dow Chemical Co.	Fe + Mo + Re + Te oxides	1	1·25 (air)		—	380	68·5	94·6	—	[298]
Toyo Soda Manuf. Co. Ltd.	Mo:Co:Fe:Bi:Sn:In:W:O 12:10:1:1:0·4:0·3:0·3:56	5	60 (air)		35 %	350		81·5	—	[299]
Mitsubishi Petro-chemical Co. Ltd.	Prepared from Mo, Ni, Fe, Bi, Co, Si, Sn and/or B compds/α–Al$_2$O$_3$	4·5	53 (air)		42·5			77·2	—	[300]
Nitto Chem. Industry Co. Ltd.	U$_{10}$Sb$_{50}$W$_{0·5}$Te$_{1·0}$O$_{130·2}$(SiO$_2$)$_{60}$	—	—		—	460	93·0	68·0	—	[301]
Mitsubishi Rayon Co. Ltd.	P + W + Mo + Te + O	6	12	52	30	340	96·0	85·0	—	[302]
Standard Oil Co. (Ohio)	Mg$_{4·5}$Fe$_4$Bi$_2$P$_{0·5}$Mo$_{12}$O$_{51}$	—	—		—	360	78·7% C to acrolein			[303]
Celanese Corp.	Co$_{83}$Mo$_{100}$Te$_{0·65}$O$_{467}$	1	36		9·4	350	46% prod. cont'g 50% acrol.			[304]

† More specific and active than Bi + Mo alone
a = 1–25, b = 1, c = corres. to Sb. Var. eg. Sb/Fe = 8:8:1.

contains a Co–Mo–Bi oxide catalyst, and is used for the further oxidation of acrolein to acrylic acid. The overall selectivity towards acrylic acid is about 40–50 %.

A similar two-reactor system, giving superior yields of acrylic acid at lower temperatures, has been developed by BASF [316]. The first reactor, which converts a propylene–air–steam mixture (120:1440:660 l/h) to acrolein, contains a catalyst consisting of a Mo–W–Te oxide mixture supported on steatite. The activity of the catalyst is progressively increased from the reactor to its exit by increasing its loading. The acrolein stream from the first converter then enters a similar reactor at 255°C, containing a Mo–W–V–Fe/steatite catalyst, also of increasing activity. In this process, 60 mole % acrylic acid is obtained at 94 mole % propylene conversion.

The direct air oxidation of propylene to acrylic acid would clearly be desirable and although this reaction is claimed to occur [317], it is often accompanied by the formation of large quantities of acrolein. Table 14 describes some fairly recent patents which claim to achieve the desired oxidation.

B. Kinetics

Comparatively few studies of this process have been reported in the literature, although Baryshevskaya et al. [348] appear to have investigated Sn–Mo and Co–Mo–Bi catalysts in some detail. Kholyavenko et al. [348] initially investigated the oxidation of propylene over mixed Sn–Mo oxides and reported that the oxidation started at 200°C and yielded, apart from acrylic acid, acetaldehyde, acrolein, carbon oxides and traces of propionic acid. As the temperature was increased, so the yields of acrylic and acetic acids increased, until a certain optimum temperature was attained (340°C for acrylic acid and 380°C for acetic acid). Thereafter, the yields decreased, with a corresponding increase in the amounts of acrolein, carbon monoxide and carbon dioxide. This phenomenon was attributed to decomposition of the acids in the catalyst pores, and so the effect of various catalyst supports (porcelain, corundum and carborundum) was also examined. With supported catalysts, oxidation began at 300–320°C and again the yields of acrylic and acetic acids increased with increasing temperature. The yield of acetic acid passed through a maximum until, at 460–500°C, the amount of acrylic acid formed was very much greater than that of acetic acid. The highest yields of acrylic acid and acrolein were reported using porcelain-supported catalysts and the lowest, using porous corundum-supported catalysts. Baryshevskaya and Kholyavenko [349, 350] have also reported the results of investigations of the kinetics of the process and of the effect of catalyst preparation on the oxidation.

Co–Mo–Bi oxide catalysts have been more extensively examined. Baryshevskaya and co-workers [351], for instance, oxidized propylene in the presence of steam, over such catalysts at 420–480°C. Activation energies were

Table 14. Some commercial catalyst configurations for the oxidation of propylene to acrylic acid

Authors	Catalyst composition	Feed-gas composition				T(°C)	C(%)	Y(%)	S(%)	Ref.
		C_3H_6	O_2	N_2	steam					
Asahi Electro-Chem. Co. Ltd.	Mo + Co + W + Te + Sb (15:12:3:1:2)	1·0-1·5	—	—	6·0	420	—	—	59·4 ac. ac. 3·9 acrol.	[337]
Gulf Res. and Dev. Co.	Co molybdate + $MoTe_2O_5$	—	—	—	—	435	58	58·6 ac. ac. 28·8 acrol. 1·6 acetic 11·0 CO_2		[338]
Gulf Res. and Dev. Co.	Co molybdate + WTe_2	11	55 (air)	—	34	435	54·9		51·7 ac. ac. 33·9 acrol. 2·6 acetic 11·8 CO_2	[339]
National Distillers and Chem. Corp.	2% Pd + 30% H_3PO_4/Al_2O_3	67	48 (air)	—		150	—	50·4 ac. ac.	83·6 ac. ac.	[340]
Knapsack A.-G.	2-layer metal oxide system 1st layer = Ag–Fe–Bi–P–Mo–O (0:12:1:0·3:12:38:41:—) 2nd layer = Co–Mo–Te–B–O on SiC or SiO_2	—	—	—	—	390	—	69·8 ac. ac.	—	[341]
Asahi Glass Co. Ltd.	Co_5 Si Mo_{12} Te_2 O_{47}	6	12	32	50 %	400	98·9	—	49·3 ac. ac.	[342]
Asahi Electrochemical Co. Ltd.	Mo (15%) Co (12%) W (3%) Te (1%) Fe (3%) O	—	—	—	—	430	—	—	61·7 ac. ac. 4·9 acrol.	[343]
Toyo Soda Manufac. Co. Ltd.	2·0% P, 47·2% Mo, 43·3% Co, 3·2% Fe, 4·3% Te oxides	2	63 (air)	—	35 mole %	370	—	35% ac. ac. 34% acrol.	—	[344]
Jakubowicz et al.	Co_a Fe_b Mo_c Te_d Si_e O_f a = 1–11·8, b = 0·02–5, C = 12, d = 0·02-4, e = 1–80, f = 37–180	7·5	52·5 (air)	—	40 %	370	—	60·2 ac. ac.	—	[345]
Mitsubishi Petro-chemical Co. Ltd.	Mo:Sn:P:Te:Fe:Si 12:1:4:1:1:5·9	—	—	—	—	425	—	47·1 ac. ac. 37·7 acrol.	—	[346]
Eastman Kodak Co.	As_2O_3 + Nb_2O_5 + $Ce(MoO_4)_3$ + $Si(MoO_4)_2$ + $Cr_2(MoO_4)_3$	—	—	—	—	—	14·5	43·3 ac. ac.	—	[347]
B.A.S.F. A.-G.	Co + Mo + Te	1	8 (air)	—	9	380	86	9 acetic 41 ac. ac. 35 acrol. 20 ac. ac.	—	[318]
Distillers Co. Ltd.	V + Sb (2·5 to 3·5:1) (+ Fe, Sn, Cu)	50%	50% (air)	—	45%	355	—	—	—	[319]

Author	Catalyst†		(air)		Temp	%	Products	Ref
Eden	$Mn + Mo + Te + P + O$	1	3·0	4·06	390	99	34 ac. ac.	[320]
B.F. Goodrich and Co.	$MoO_3 + TeO_2 +$ metal phosphate	1	molar vols. 2·96 (air)	4·36	372	99	35 acrol.	[321]
W.R. Grace and Co.	$Co + Mo + Bi$ (9:10:1)	7·5	37·5 (air)	55	426	35	46 ac. ac.	[322]
Hirota et al.	$Mo + V + Bi$ (0·8:1:0·2)	15	4·8 11·5	68·7	400	12	34 acrol. 67 ac. ac. 20 acrol. 48 ac. ac.	[323]
I.C.I. Ltd.	$Mo + Te + P$ (1:0·2–0·75:0·5)		—		370 to 423	—	15–34 acrol. ca. 25 ac. ac.	[324]
Japan Catalytic. Chem. Ind.	$W + Sn + Co + Mo + Te$	1	60 (air)	39	330	—	7 acrol. 62 ac. ac.	[325]
Nakano et al.	$Mo + Sb + V + Te$ (2:7:2:1)	3·2	64·5 (air)	32·3	375	85	40 ac. ac.	[326]
Kashiwabara and Nakamura	$Mo + Sn + V + Te + W$ (5:2:1:0·5:1·5)	1	1·6	6	450	—	67 ac. ac.	[327]
Eden	$3MoO_3 + TeO_2 + ThP_2O_7$	1	3	4·1	345	—	40 acrol. 27 ac. ac.	[328]
Same	Same	Same	—	—	425	—	32 acrol. 45 ac. ac.	[328]
Trapasso and Wenrick	$Ni + Cr + Te + Mo + Si$ 1:1:1:4:2·5	0·032	0·08	0·127 mol/h	425	95	23 acrol. 62 ac. ac.	[329]
	1:1:2:4:2·5	Same	Same	—	430	67	10 acrol. 83 ac. ac.	[329]
Rohm. and Haas Co.	Co molybd. $+$ 0·1 to 1 w% Cu_2Te	1	4·6 (air)	4	438	67	33 acrol. 43 ac. ac.	[330]
Scherhag et al.	P_2O_5 (63%) $+$ Cu/Bi (3·7%)	1	8·25 (air)	6	440 to 460	45	40 to 42 ac. ac. 47 acrol.	[331]
S.N. des Pétroles d'Aquitaine	NH_4 phosphovanadomolybdate (35%) + telluric acid (35%)/silica	1	1		440	66	28 ac. ac.	[332]
Young and Reynolds	Nb_2O_5 (9·5) $+ MoO_3$ (20·5)/70% porous $SiO_2(As_2O_3$ (g) added)	—	—				17 to 26% → 45–61 ac. ac. 5 to 18 → 13–18 AA	[333]
B.F. Goodrich and Co.	MoO_3 (100) $+ TeO_2$(33·25) $+ Mn_2P_2O_7$ (33·25)	1	2·95	4·16	378	93	40 acrol. 47 ac. ac.	[334]
Parthasarthy et al.	$Co + Bi + Mo$ (0·9:0·1:1·0)	7·5	37·5	55	798°F	35	48 ac. ac. 20 acrol.	[335]
Jakubowicz et al.	$Co_{11·25} Mo_{12} O_{47·25}$	7·5	52·5 (air)	40·0	380	—	54 ac. ac.	[336]

Contact times, although variable, are often long (20 to 40 s in some cases).
† = usually as oxides. ac. ac. = acrylic acid; acrol. = acrolein; AA = acetaldehyde; acetic = acetic acid.

derived for the formation of acrylic acid (21 kcal. mole^{-1}), acrolein (16 kcal. mole^{-1}), carbon monoxide (30 kcal. mole^{-1}) and carbon dioxide (40 kcal. mole^{-1}). Over a wide range of concentrations, the formation of both acrolein and acrylic acid was independent of oxygen partial pressure. Increasing the amount of steam in the feed-gases increased the formation of acrylic acid and decreased the formation of the carbon oxides, whilst an increase in propylene concentration increased the formation of all oxidation products. Further investigation [352] enabled the following relationships to be obtained:

Rate of formation of acrylic acid $=K[C_3H_6]^{1 \cdot 2}[H_2O]^{0 \cdot 3}/(1 + kC)$

Rate of formation of acrolein $= K'[C_3H_6]^{1 \cdot 2}[H_2O]^{0 \cdot 1}/(1 + kC)$

Rate of formation of carbon oxides $= K''[C_3H_6]^{1 \cdot 5}[O_2]^{0 \cdot 5}/$
$$(1 + k'C)[H_2O]^{0 \cdot 1}$$

where K, K' and K'' are the respective rate-constants, k and k' are temperature-dependent constants and C = [acrylic acid] + [acrolein].

The mechanism of the oxidation was said to be complex, involving parallel and consecutive reactions and it was found that the amount of acrylic acid formed directly from propylene was about the same as that produced from acrolein within the system. Addition of acrolein, carbon monoxide and carbon dioxide did not affect the rate of propylene oxidation. Baryshevskaya *et al.* [353] have also studied the oxidation of acrolein to acrylic acid over Co–Bi–Mo catalysts. It was reported that the rate of oxidation of acrolein to acrylic acid was some 15–25 times greater than the rate of the direct oxidation of propylene to acrylic acid. The rate of acrolein oxidation was found to be independent of the oxygen concentration and addition of steam to the reacting gases enhanced the formation of acrylic acid at the expense of the carbon oxides.

Finally, Suvorov *et al.* [354] studied propylene oxidation in the presence of vanadia–alumina catalysts in the temperature range 240–320°C. Oxidation products included acrylic, propionic, acetic and formic acids, acrolein, acetaldehyde and formaldehyde. This catalyst is poor, however, giving a low yield of acrylic acid (16·5% at 240–260°C) which falls as the temperature increases.

From the foregoing discussion and Table 14, it can be seen that catalysts such as Co–Bi–Mo oxides, have the disadvantage that acrolein is produced concurrently with acrylic acid. Recently, however, Campbell *et al.* [355] have described the development of catalysts, in the presence of which, propylene may be oxidized to acrylic acid with only traces of acrolein as a side-product. Acetic acid, is, however, now the significant by-product. These catalysts contain 5–10% As_2O_5, 10% Nb_2O_5 (or Ta_2O_5) and 20% MoO_3, supported on silica. At 400°C with a propylene-air-steam feed, conversions to acrylic acid of at least 20 mole% are obtained in a single pass, the yield of acrylic acid being about 50 mole%. Unfortunately, such catalysts tend to lose arsenic as As_2O_3

during service, but by adding As_2O_3 to the feed gases, a constant catalytic activity can be maintained for over 1100 hours.

3.4. To Acrylonitrile

A. Processes

Acrylonitrile is an extremely important compound in the petrochemical industry, and recently several closely related reviews have appeared in the literature which summarize the commercial production and uses of the substance and also examine present and future markets [356–361]. In this review, the commercial production of acrylonitrile will not be dealt with in any detail. Suffice it to say that Stobaugh *et al.* [356] predict that the fluidized-bed ammoxidation of propylene, discovered and developed by Standard Oil (Ohio) [246, 362–364] and employing Bi–Mo–O or U–Sb–O catalysts, is likely to become the generally accepted process for acrylonitrile production; there appears to be some justification for this claim [365].

Table 15 gives a list of recent patent claims for the ammoxidation of propylene.

B. Kinetics and Mechanism

(a) Bi_2O_3–MoO_3 Catalysts. Idol [362] discovered that, when a mixture of air, propylene and ammonia was passed through a fluidized bed of P-promoted bismuth molybdate at 470°C, almost all the propylene reacted, and 65% appeared in the products as acrylonitrile. Other products formed included methyl cyanide, hydrogen cyanide, and carbon monoxide and dioxide. Acrolein was a trace product only. A very much later investigation by Cathala and Germain [393] has revealed that, with bismuth molybdate catalysts, the selectivity of the process towards acrylonitrile passes through a maximum at about 440°C (Table 16), and activation energies for the formation of acrylonitrile (17 kcal. mole^{-1}), acrolein (38 kcal. mole^{-1}), methyl cyanide (11 kcal. mole^{-1}), ethylene (25 kcal. mole^{-1}) and carbon oxides (40 kcal. mole^{-1}), during the oxidation, were also obtained.

There is a great similarity in the overall features of the oxidation and ammoxidation of propylene over bismuth molybdate catalysts; ammoxidation is first-order with respect to propylene and zero-order with respect to both oxygen and ammonia [182, 184, 189, 394, 395]. Several detailed studies have been made of the mechanism of both the oxidation and ammoxidation of propylene. Thus Gel'bshtein and co-workers [396] studied catalysts of various compositions, including MoO_3, $Bi_2O_3 \cdot 3MoO_3$, $Bi_2O_3 \cdot 2MoO_3$, Bi_2O_3, and several having a Bi/Mo atomic ratio greater than one. Catalysts containing the largest amounts of the compound $Bi_2O_3 \cdot 2MoO_3$, were found to have the greatest activity in both reactions whilst MoO_3 had a low selectivity and activity

Table 15. Some commercial catalysts for the ammoxidation of propylene to acrylonitrile

Authors	Catalyst Composition	Feed-gas composition				$T(°C)$	$C(\%)$	$Y(\%)$	$S(\%)$	Ref.
		C_3H_6	NH_3	steam	air					
Ube Industries Ltd.	$BiSbO_4$–$BaMo_3O_{10}$	1	1	2	7·5	470	47·4	—	90·4	[366]
Monsanto Co.	Sb_2O_3 + UO_2 on Al_2O_3†	7·4	8·5	17·7 (O_2)	66·4 (He)	500	26·9	—	75·3	[367]
Ube Industries Ltd.	Mo (20%) + Bi (45%) + Sb (35 at %) O	1	1	1	5	470	49·2	—	85·4	[368]
Ube Industries Ltd.	$BiSbO_4$ + WO_3 (75:25)	1	1	2	7·5	—	77·5	68·2	90·2	[369]
Asahi Chem. Ind. Co. Ltd.	P:Mo:Bi:Co:Fe (1:12:3:5:1)	—	—	—	—	450	92·2	—	—	[370]
Asahi Chem. Ind. Co. Ltd.	MoO_3 (24-42%), Bi_2O_3 (14·82%), Fe_2O_3 (5·08%), CoO (4·24%), Na_2O (0·44%), P_2O_5 (1·01%), SiO_2 (50%)	35·5	44·5	—	340 (1/h)	480	97·2	76·4	—	[371]
Nitto Chem. Ind. Co. Ltd.	Fe–Sb–Si	8·0	9·3	—	82·7 v %	450	91·5	70·8	77·0	[372]
Ube Industries Ltd.	Mo–Bi–Sb (32·5:50:17·5)	1	1	1	5	470	59·7	91·2	—	[373]
	$BiSbO_4$–SnO_2–$Bi_2(MoO_4)_3$ (60:15:25)	1	1	1	7	470	59·3‡	—	90·6§	[374]
SNAM Progetti S.p.A.	W + Bi (or Te) + Fe oxides	1	1·2	—	10 (O_2)	480	70·0	—	68·0	[375]
Knapsack A.-G.	Fe + Bi + Mo + P (a:10:15:0·2)	1	10	—	1·25	450-490	91·0¶ 96·0‖	—	78·0¶ 74·0‖	[376]
Montecatini Edison S.p.A	Te + Ce + Mo + O (4:5:12:54)/ SiO_2	—	—	—	—	440	84·0	65·0	—	[377]
Erdoelchemie G.m.b.H.	Bi (6-26), Mo (2-20), Fe (0·5-10), P (0·1-1·0%) oxides	1	1-1·1	—	9·5-10	456-460	—	73·0	—	[378]
SNAM Progetti S.p.A.	U–Te oxide (1:4) + 25% SiO_2	—	—	—	41%	—	84·3	—	69·4	[379]
S.N. des Pétrol. d'Aquitaine	U–Te–Mo oxide (1:2:0·5 at.rat.)	3·4	4·4	51·2	—	450	83·0	—	78·0	[380]
	Fe + Mo + Te/SiO_2	—	—	—	—	—	73·0	—	84·0	[381]
Monsanto Co.	$Bi_9PMo_{12}O_{52}$/SiO_2	1 (C_3H_8)	1·2	—	12	500	55·9	62·3	—	[382]
Standard Oil Co. (Ohio)	82·5:17·5% $Ni_{10·5}FeBiPMo_{12}O_{57}$–$SiO_2$	1	1·5	—	11	400	78·4	—	—	[383]

Company	Catalyst					Temp. (°C)				Ref.
SNAM Progetti S.p.A.	U–Te oxide (1:4) + 25–50% SiO_2	1	1·1	5	12	430–460	—	53·5	63·4	[384]
Nitto Chem. Ind. Co. Ltd.	$U_{10}Sb_{50}W_{0.5}Te_{1.0}O_{130.2}(SiO_2)_{60}$	—	—	—	—	480	93·0	77·0	—	[385]
S.N. des Pétrol d'Aquitaine	Fe:Sb:W:Te:Si (10:25:0.5:30)	—	1	—	—	450	—	—	78·0	[386]
	Fe + Mo + Te	1	—	—	2 (O_2)	415	—	71·5	88·0	[387]
Ube Industries Ltd.	75:25 wt. ratio {$BiSbO_4$–SnO_2 mixture 80:20 wt. ratio}–$Bi_2(MoO_4)_3$	—	—	—	—	470	—	—	90·6	[388]
Monsanto Co.	$Sb+U+Ni+V$ (5:1:0·5:0·5)/SiO_2	1 (C_3H_8)	1·2	—	12 (O_2)	—	—	63·5	—	[389]
Nitto Chem. Ind. Co. Ltd.	$Fe_{10}Sb_{25}W_{0.25}Te_{0.5}O_6(SiO_2)_{30}$	1	1·3	—	2·2 (O_2)	440	—	—	78·0	[390]
Aykan	$PbTi_3O_7$ (S = 10·4 m^2 g^{-1})	25·2%	7·2% (NO)	—	67·6% N_2	510	—	72·0	—	[391]

† Sb–U–O cats. are regenerated by treating 0·1–24 h with N_2 at 427–982°C. Regenerated cat. gave C_3H_6 conversion = 96·8%; selectivity = 83·5%; S = 19·6 m^2 g^{-1}. (Cf. 88·4%; 73·1% and 20·1 m^2 g^{-1} for exhausted cat. and 92·6%; 81·5% and 20·2 m^2 g^{-1} for new cat.)—Ref. [392].
‡ Cf. 25·5% for $BiSbO_4$ alone and 40·2% for SnO_2 alone.
§ Cf. 10·9% for $BiSbO_4$ alone and 10·3% for SnO_2 alone.
¶ a = 1·5; s = 21 m^2 g^{-1}
‖ a = 12; s = 5 m^2 g^{-1}.

Table 16. Selectivity towards acrylonitrile during the
ammoxidation of propylene over bismuth molybdate at
various temperatures

Temp. (°C)	384	422	436	462	481
Selectivity (%)	82	84	84	74	67

and Bi_2O_3 was completely inactive. It was found, however, that, whereas both oxidation and ammoxidation were first-order in propylene and zero-order in oxygen, the order with respect to ammonia was zero only at temperatures in excess of 440°C. The overall rate of the ammoxidation reaction could, it was found, be expressed by the equation:

$$\text{Rate} = 3 \cdot 8 \times 10^5 \exp\left(-16,000/RT\right)[C_3H_6]$$

whilst a similar expression for the oxidation had the form:

$$\text{Rate} = 1 \cdot 0 \times 10^5 \exp\left(-15,500/RT\right)[C_3H_6]$$

The great similarity between the activation energies is consistent with the view that the rate-limiting step is common to both reactions. This step was thought to involve the dissociative chemisorption of propylene, giving an allyl species and a hydrogen atom, on a doublet site involving either $Bi^{3+} + Mo^{6+}$ cations or two molybdenum cations. The activation energy of oxygen adsorption in the temperature range 400–450°C, was also measured and, for MoO_3, $Bi_2O_3 \cdot 2MoO_3$ and Bi_2O_3, the values obtained were 20, 16 and 14 kcal. mole^{-1} respectively. Similarly, the heat of oxygen adsorption decreased in the same order (8·5, 5·8 and 3·1 kcal. mole^{-1}, respectively).

Earlier, Kolchin et al. [35] had determined the specific rate constants for the formation of the major products in both the oxidation and ammoxidation of propylene over the α-, β- and γ-phases of bismuth molybdate:

Phase Comp.	$k_{sp.}^a$ (min^{-1} m^{-2} × 10^{-2})		S^y	$k_{sp.}^b$ (min^{-1} m^{-2} × 10^{-2})			S^y
	Acrolein	CO_2		Acrylonitrile	CO_2	HCN	
$Bi_2O_3 \cdot 3MoO_3$ (α)	5·29	1·85	74	14·1	5·24	2·50	65
$Bi_2O_3 \cdot 2MoO_3$ (β)	9·4	1·38	87	16·6	3·16	1·28	76
$B_2O_3 \cdot MoO_3$ (γ)	0·55	0·5	52	2·3	1·14	1·13	50

$$^a = 475°C \qquad ^b = 480°C$$

Roughly the same activity relationship between the various phases was found in both oxidation and ammoxidation. The activation energies for the formation of both acrolein and acrylonitrile were reported to be almost identical at 19 and 21 kcal. mole^{-1} respectively, and were independent of the phase of the bismuth molybdate. (Margolis [397] has also investigated activity in $Bi_2O_3 \cdot MoO_3$-promoted ammoxidation as a function of the phase. From the results it was concluded that a doublet-centre mechanism (cf. Ref. 396) for the reaction could not be correct, since the γ'-phase of bismuth molybdate, which has both Bi^{3+} and Mo^{6+} ions in one plane, is less active and selective than the γ-phase which only has Mo^{6+} ions.) Similarly, Ohdan et al. [398] ammoxidized propylene in the presence of bismuth molybdate, and like Kolchin [35], concluded that the α- and β-phases were most active in the reaction. Infra-red studies also led them to conclude that the catalytically-active species was an oxygen atom, doubly-bonded to molybdenum, having characteristic absorptions at 910, 940 and 955 cm^{-1}. Beres et al. [204], in an investigation of the influence of catalyst preparation on the subsequent activity of Bi–Mo–O catalysts, found that catalysts containing the β-phase only are characterized by a very high activity in the synthesis of acrylonitrile (more than 90% of the propylene was converted, at 490°C, to acrylonitrile, with a selectivity of 80%). Although the Bi/Mo ratio of the β-phase is unity, Beres et al. observed that precipitation from solutions with this ratio yielded a mixture of both α- and β-phases. Pure β-phase could, however, be obtained by complexing part of the molybdenum in solution as magnesium molybdate before precipitation, thus increasing the Bi/Mo ratio. Dalin et al. [399] have recently reported the results of an interesting study of the changes occurring in silica-supported bismuth molybdate catalysts associated with their prolonged use in propylene ammoxidation at 470°C. The changes are characterized by a fall in the rate of acrylonitrile formation and a lowering of the activation energy of the reaction. As made, catalysts with a Bi/Mo ratio in the range 0·4–2·0 contained $Bi_2O_3 \cdot MoO_3(\gamma)$, $Bi_2O_3 \cdot 2MoO_3(\beta)$ and $Bi_2O_3 \cdot 3MoO_3(\alpha)$. With use, it was found that the highly active α- amd β-phases gradually changed to the less active γ-phase, and after 900 hours, the catalyst consisted of $Bi_2O_3 \cdot MoO_3$ with MoO_3 in solid solution. The MoO_3 initially formed was amorphous but slowly crystallized from the solid solution. Loss of catalytic activity was thus ascribed, not only to the conversion of the α- and β-phases into the γ-phase, but also to crystallization of MoO_3.

The mechanism of the formation of acrolein from propylene, studied by Adams and Jennings [177, 191] has been discussed (Section 3.2.), and it is the opinion of Callahan et al. [182] that the formation of acrylonitrile takes place by a similar mechanism:

$$CH_3—CH{=}CH_2 \xrightarrow{(1)} H_2C \cdots CH \cdots CH_2 \xrightarrow{(2)}$$

$$H_2C{=}CH—CH \xrightarrow{(3)} H_2C{=}CH—C$$

acrolein   acrylonitrile

where (1), (2) and (3) represent the first, second and third hydrogen-abstractions. Although this scheme is acceptable in part, some questions still remain unresolved. For example, the role of acrolein in acrylonitrile formation is still uncertain. Kinetic studies at 425°C, by Callahan *et al.* of the ammoxidation of both propylene and acrolein showed that the activation energy for propylene ammoxidation (17–19 kcal. mole^{-1}) was higher than that for the ammoxidation of acrolein (7 kcal. mole^{-1}). It was found that the data could be adequately described by the following triangular reaction scheme:

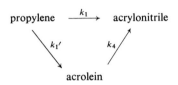

where $k_1/k_1'/k_4 = 100/2·5/200$ and k_1', k_4 and k_1 are 0·005, 0·4 and 0·195 s^{-1} respectively. Cathala and Germain [400] have reported very similar values for $k,/k_1'/k_4$ (100/3·3/535) in a study of propylene ammoxidation at 460°C over quartz-supported bismuth molybdate according to a similar triangular scheme. Shelstad and Chong [401] have, however, proposed the linear scheme

propylene $\xrightarrow{k_1'}$ acrolein $\xrightarrow{k_4}$ acrylonitrile

$\downarrow k_6$

CO, CO$_2$

in which $k_1'/k_4 = 1/16$ [401].

A fuller reaction scheme proposed by Cathala and Germain [393], again showed the excellent agreement in rate constants with those of Callahan [182]. This scheme and the rate data are shown below:

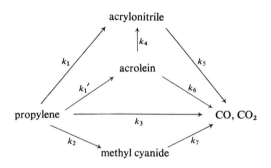

Rate constants

k_1	k_1'	k_2	k_3	k_4	k_5	k_6	k_7	T°C	Ref.
100	3·3	3·3	2	535	18	0	37	460	[291]
100	2·5	5	8	200	—	—	36	425	[141]

Both Callahan *et al.* [182] and Cathala and Germain [393] have investigated the ammoxidation of acrolein in order to elucidate its role in propylene ammoxidation. Callahan *et al.* [182] concluded that, in the case of propylene, acrylonitrile was formed largely by a mechanism "not involving acrolein as a vapour-phase intermediate". Cathala and Germain, although not reaching any definite conclusion, derived a reaction scheme, with rate constants, which accounted for most of the products encountered in the ammoxidation of both propylene and acrolein:

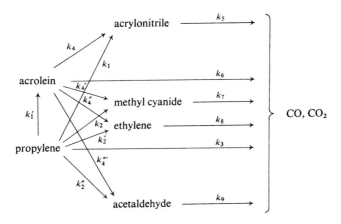

Products such as methyl cyanide, hydrogen cyanide, ethylene and the oxides of carbon were secondary products, and certain of these products (e.g. ethylene) were formed by homogeneous gas-phase reactions of radicals, such as $CH_2=CH-\dot{C}O$, derived from acrolein. A summary of the rate constants obtained by Cathala and Germain for propylene ammoxidation and various related reactions is given in Table 17.

The precise role of ammonia within the ammoxidation system is still not known, although attempts have been made to elucidate it. For example, Seeboth *et al.* [402], after an initial investigation of propylene ammoxidation in the presence of a typical Standard Oil catalyst ($Bi_9PMo_{12}O_{52}$), examined the function of ammonia using a silica-supported catalyst of formula $Bi_6Fe_{12}PMo_{12}O_{65.5}$. The reaction was investigated at two temperatures (375 and 425°C) and it was found that a 60% yield of acrylonitrile was obtained at

Table 17

Reaction	k	k_1	k_1'	k_2	k_2'	k_2''	k_3	k_4	k_4'	k_4''	k_4'''	k_5	k_6	k_7	k_8	k_9
Propylene ammox.	5·2	4·8	0·16	0·16	—	0	0·1	25·8	—	—	0	0·9	0	1·77	—	—
Propylene ox.	8·0	—	5·85	—	0	0·06	2·0	—	—	—	—	—	6·2	—	—	—
Acrolein ammox.	170·0	—	—	—	—	—	—	145·0	1	3	0	28	22	—	—	—
Acrolein ox.	97·0	—	—	—	—	—	—	—	—	7	42	—	48	—	0	390

425°C compared with one of only 27% at 375°C, and at 425°C, the nitrogen-content of the reaction products corresponded to the amount of ammonia converted. An assumption was made that, if acrylonitrile was produced via some intermediate such as acrolein or acetaldehyde rather than propylene, then the yield would be independent of the reaction temperature. In the light of the experimental results, however, two mechanisms were proposed for the formation of acrylonitrile. One mechanism, said to operate at temperatures above 400°C, involved simultaneous dehydrogenation of ammonia, to form species such as NH_2 or NH, and formation of either an allyl or a carbene (CH_2=CH—CH:) intermediate from propylene (the latter being the rate-determining step), the carbene or allyl species then reacting with NH_2 or NH to give acrylonitrile. At temperatures below 400°C, ammonia dehydrogenation was thought not to occur, acrylonitrile being formed in a direct reaction of ammonia with acrolein.

An alternative mechanism involves an oxide of nitrogen (such as nitric oxide) rather than ammonia, as the species responsible for ammoxidation. It is known that the oxidation of ammonia over catalysts such as bismuth molyb-date yields oxides of nitrogen such as N_2O and NO [403]. The formation of acrylonitrile by the direct reaction of propylene and nitric oxide over a variety of catalysts is also well-documented [405–408]. It is possible, therefore, that acrylonitrile may be formed by the reaction of nitric oxide with either acrolein or propylene. In the presence of oxygen, however, the reaction of nitric oxide with propylene is probably more than ten times slower than the reaction of nitric oxide with oxygen and the proposed reaction therefore seems unlikely. A very attractive theory for the role of ammonia in acrylonitrile formation has, however, been proposed by Schuit [404] and this is based on some observations made by Giordano [403]. It had been observed that, in the presence of bismuth molybdate catalysts (Bi/Mo = 0·3 to 0·7) ammonia was oxidized to nitrogen gas and nitrous oxide, whereas in the presence of catalysts having a Bi/Mo ratio between 0·7 and 1·0, nitric oxide was the main product. Propylene was found to inhibit the reaction strongly, acrylonitrile becoming the preferred product. Schuit interpreted these observations in terms of Matsuura's theory [211] that certain sites (A-sites and B-sites) exist on the surface of bismuth molybdate catalysts. The work of Matsuura will be examined in detail in Chapter 4, Section 4.2 but briefly, it was proposed that the A-sites (present in low concentration on the surface) comprise an O^{2-} ion in a special position and A-site vacancies (formed by partial reduction of the surface) were able to adsorb both oxygen and water. B-sites were found to adsorb gases such as propene, but-1-ene, cis-but-2-ene and buta-1,3-diene and, it was proposed, consist of pairs of surface O^{2-} ions different from those encountered in A-sites; B-sites it was assumed are present in greater concentration. Since ammonia is similar to water, Schuit [404] assumed that it was similarly adsorbed, i.e. on an anion-vacancy (V_A) left on an A-site. Back-migration of a hydrogen atom from there to oxygen in a B-site and thence to an A-site, followed by desorption of water, leads to the situation where nitrogen atoms (N_A) are situated on A-sites. In the absence of propylene, recombination:

$$2N_A \rightarrow N_2$$

or reaction with an A-site oxygen atom:

$$2N_A + O_A \rightarrow N_2O$$

occurs, but in the presence of propylene, N_A reacts with adsorbed allyl to give acrylonitrile.

Finally, in a discussion of the catalytic properties of bismuth molybdate, mention must be made of a study by Aykan [409] on the reduction of silica-supported Bi–Mo–P oxide catalysts. Ammoxidation of propylene was achieved in the absence of oxygen (Feed: $5.1\% C_3H_6$, $5.1\% NH_3$, $89.8\% N_2$) and changes in the catalyst were examined by X-ray diffraction. It was found that, on reduction, the catalytically active compound $Bi_2O_3 \cdot 3MoO_3$ was transformed ultimately to metallic bismuth and MoO_2, via koechlinite ($Bi_2O_3 \cdot MoO_3$), according to the equations:

$$Bi_2O_3 \cdot 3MoO_3 - (C_3H_6, NH_3) \rightarrow Bi_2O_3 \cdot MoO_3 + 2MoO_2 + O_2$$

$$Bi_2O_3 \cdot MoO_3 - (C_3H_6, NH_3) \rightarrow 2Bi + MoO_2 + O_2$$

As the catalyst became increasingly reduced, so its activity rapidly declined, the phase-changes within the catalyst and product formation corresponding quantitatively to the depletion of lattice oxygen. It was concluded, therefore, that under normal conditions of ammoxidation, simultaneous reduction and reoxidation of the catalyst occurred.

(b) UO_3–Sb_2O_5 Catalysts. The discovery and exploitation of catalysts containing complex uranium and antimony oxides is again due to Standard Oil (Ohio) [246] and most of the fundamental studies of the nature of the catalytically active phases within this catalyst and the reaction mechanism, have been carried out by Sohio personnel [249–251]. Structural studies have however also been performed by Aykan and Sleight [410].

Initially, Grasselli and Callahan [249] reported that two crystalline phases—$(UO_2)Sb_3O_7$ (phase I) and $Sb_3U_3O_{14}$ (phase II)—existed in UO_3–Sb_2O_5 catalysts. Later work [250] indicated that phase I is better described by the formula USb_3O_{10} and phase II by the formula $USbO_5$. It was found that the compound, USb_3O_{10}, is the active and selective phase in both propylene oxidation and ammoxidation over the catalysts, but $USbO_5$, although equally active, is less selective. This conclusion was also reached by Nozaki and Okada [411]. Like bismuth molybdate, the two compounds can be represented as layer structures; these are depicted below (only the metal atoms are shown):

$$USb_3O_{10} \; (Phase \; I) \qquad\qquad USbO_5 \; (Phase \; II)$$

Orthorhombic. Space group Fddd-D_{2h}^{24}

$a = 7.346$ Å, $b = 12.72$ Å, $c = 15.40$ Å, $Z = 8$

U	Sb	U		U	U	U
	Sb	Sb			Sb	Sb
Sb	U	Sb		U	U	U
	Sb	Sb			Sb	Sb
U	Sb	U		U	U	U

Pure phase I is prepared from uranium-antimony oxides (U/Sb = 1:9·2) by dissolving the excess antimony oxide in hydrochloric acid. Phase II is prepared by thermal decomposition of Phase I at 1090°C. In these compounds, it is believed [251] that the active site involves the grouping:

$\square$ = vacancy

This grouping, it is proposed, is generated during preparation of the catalyst according to the scheme:

During oxidation, adsorption of propylene as a π-allyl complex is said to occur:

After this, depending on the conditions, oxidation or ammoxidation will occur:

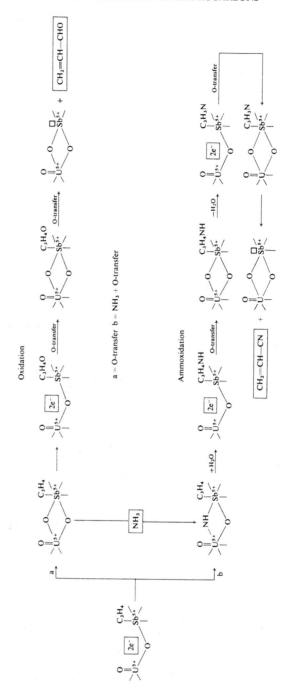

(c) *Other Mixed Oxide Catalysts.* Ohdan and co-workers have examined the efficiency of the ternary system Bi–Mo–Sb in the oxidation and ammoxidation of propylene [253–255, 412]. Catalyst preparation and the nature of the compounds formed therein, has already been dealt with in Section 3.2. Catalysts consisting of $Bi_2O_3 \cdot 3MoO_3$ and $Bi_2O_3 \cdot Sb_2O_5$ in the mole ratio 23:77 and 64:36 were found to be more active and selective for ammoxidation than the binary catalysts. Thus, at 470°C with a propylene:ammonia:steam:air feed (1:1:1:5), 40% propylene was converted with a selectivity towards acrylonitrile of over 80%. Kinetic studies revealed reaction orders very different from those found with the binary Bi_2O_3–MoO_3 system. For example, apparent reaction orders with respect to propylene, ammonia and oxygen were found to be 0·76, −0·76 and 0·49 respectively. The activation energy for acrylonitrile formation was determined as 13·7 kcal. mole^{-1}. Methyl cyanide, a significant by-product, was assumed to be formed via an intermediate common to acrylonitrile formation. Ohdan et al. [412] have examined the effect of steam in propylene ammoxidation at 470°C over $Bi_2O_3 \cdot Sb_2O_5$ and $Bi_2O_3 \cdot 3MoO_3$ catalysts. In the absence of steam, ammonia was completely decomposed but no acrylonitrile was formed. The presence of steam reduced ammonia decomposition and produced a concurrent increase in the yield of acrylonitrile. Studies were also made of the decomposition of acrylonitrile over the catalysts in the presence of various hydrocarbons. Hydrocarbons were effective in decreasing the rate of decomposition. In order of decreasing effectiveness the order is: propane > ethane > ethylene > butadiene ≫ propylene > cis-but-2-ene > but-1-ene. Ohdan et al. [413] have also examined tungsten oxide–bismuth oxide catalysts (9:1 to 6:4 W/Bi atomic ratio) for propylene ammoxidation. Such catalysts are, however, fairly inactive and the selectivity towards acrylonitrile is low (30%).

The system antimony oxide–iron oxide has been thoroughly investigated by Boreskov and co-workers [414] with particular reference to the influence of chemical composition on activity and selectivity. A large number of catalyst compositions were explored, from 100 Fe atomic% to 100 Sb atomic%, using X-ray diffraction. Pure α-Fe_2O_3 was found to be a very active catalyst at 425°C (rate constant for propylene conversion 220×10^7 1 m^{-2} s^{-1}) but the major product was carbon dioxide (selectivity towards $CO_2 = 93\%$ and towards acrylonitrile = 3%). However, Sb_2O_4, although much more selective towards acrylonitrile (41%), was relatively inactive (rate constant for propylene conversion, $0·014 \times 10^7$ 1 m^{-2} s^{-1}). A range of compositions was found, however, wherein the selectivity towards acrylonitrile was greater than 80%. These composition ranges were 83·4–91·9 Sb atomic% and 61·5–75·8 Sb atomic% and such catalysts were also fairly active (rate constants 3·09 to 12·1 × 10^7 1 m^{-2} s^{-1}). X-ray diffraction revealed that, in such ranges of composition, the compound $FeSbO_4$ was formed and it was to this compound that the great increase in selectivity was ascribed. Unfortunately, with certain compositions, notably 56·5–42·5 Sb atomic%, although $FeSbO_4$ was apparently the only phase present, the selectivity of the process was fairly low (36–40%). Boreskov *et al.* [415] also investigated the energy of the oxygen bond at the surface of Fe–Sb–O catalysts. From experimental results on the reduction of these catalysts with carbon monoxide or hydrogen, three types of oxygen were detected, the first having a bond dissociation energy of 48 kcal. $mole^{-1}$, the second a bond dissociation energy between 48 and 74 kcal. $mole^{-1}$ and the third a bond dissociation energy greater than 74 kcal. $mole^{-1}$. These results were unfortunately correlated with the mechanism of but-1-ene oxidation and the selectivity of Fe–Sb–O catalysts in this reaction was ascribed to their ability to form the most weakly-bonded oxygen species at a low rate.

In other studies, Skalkina *et al.* [416] have used the Mössbauer effect to examine certain Fe–Sn–O catalysts containing Sb, Cr or Mo oxides used in propylene ammoxidation. A correlation was found to exist between the Mössbauer parameters (isomer shift (δE) and quadrupole splitting (ΔE_Q)) and the activity of the catalysts. When values of ΔE_Q were low, carbon dioxide was the preferred reaction product but, as ΔE_Q increased, so did the amount of acrylonitrile formed. Trifiro *et al.* [417] have studied a variety of molybdates with respect to their activity for propylene ammoxidation. It was found that molybdates could be classified according to their infra red absorption spectra. Three such classes were distinguished:

(1) Molybdates such as Fe, Mn and Co molybdate, which were active but not selective, were found to have a strong absorption at 940–970 cm^{-1}.
(2) Molybdates such as bismuth molybdate, which were both active and selective catalysts, had a small but intense absorption at 940 cm^{-1} and a much more intense absorption in the region 920–930 cm^{-1}.
(3) Molybdates without characteristic absorptions in the region 900–1000 cm^{-1} (e.g. Ca, Pb, Tl molybdates) were found to be practically inactive.

Trifiro *et al.* [418] studied Sn–Sb–O catalysts (Sn:Sb = 5·66) in the ammoxidation of propylene at 400°C. With increasing oxygen concentration, the propylene conversion increased, reaching a maximum at about 7%; thereafter it declined, becoming independent of the oxygen concentration at about 11 mole%. In the presence of such catalysts, the oxidation of ammonia was negligible although in the reaction with acrolein, almost all the ammonia reacted to form acrylonitrile.

Tin–antimony–iron oxide (1:4:0·25) catalysts were investigated by Crozat and Germain [419] in the temperature range 415–507°C. At 455 ± 5°C, 1:10:1·2 propylene:air:ammonia mixtures yielded acrolylonitrile with an initial selectivity of 80%, the selectivity altered little as the propylene conversion increased (up to 60%). As the temperature was increased, the activation energy for acrylonitrile formation was found to fall, from 20·6 to 17·5 kcal. mole^{-1}.

Finally, an unusual route for the synthesis of acrylonitrile has been reported by Japanese workers [420–423]. This process involves an olefin, hydrogen cyanide and oxygen and is termed "oxycyanation". The general equation for the reaction is shown below:

$$\text{>C=C<}_{H} + HCN + \tfrac{1}{2}O_2 \longrightarrow \text{>C=C<}_{CN} + H_2O$$

With ethylene as the olefin, acrylonitrile is obtained; propylene gives meth-acrylonitrile and crotononitrile. Preferred catalysts contain palladium, vanadium and caesium supported on alumina and are operated at about 340°C. With an ethylene/hydrogen cyanide/oxygen/hydrogen chloride/nitrogen (50, 3, 3, 3, 41 vol. %) feed, for example, 79 mole % ethylene is converted to acrylo-nitrile with a yield of 90 mole %.

The mechanism of oxycyanation has been discussed in a series of papers by Nakajima et al. [424].

3.5. To Acetone

In 1968, Buiten [229] and Moro-oka et al. [425] reported work with certain MoO_3-containing catalysts, which indicated that, apart from suffering oxida-tion at the allylic group to give acrolein, propylene under certain conditions (low reaction temperatures and in the presence of steam) could undergo a very different type of oxidation to give acetone and acetic acid. Only at low tempera-tures (100–300°C) did oxidation to acetone occur, higher temperatures causing the reaction to shift towards allylic oxidation and subsequent C—C bond fission. Buiten's observations involved an SnO_2–MoO_3 catalyst (or more precisely, SnO_2 covered with a monolayer of MoO_3) and, although the tempera-ture was fairly high (340–350°C), a selectivity towards acetone of between 30 and 40% was found. Moro-oka [425] has investigated the Co_3O_4–MoO_3 binary system and found that catalysts having a Co/Mo ratio between 9:1 and 7:3 were selective towards acetone formation. An investigation of the effect of temperature on the oxidation showed that a Co/Mo = 9:1 catalyst yielded acetone at 195°C with a selectivity of 81%. Later investigations by Moro-oka et al. [426, 427, 428, 228] showed that other transition metal oxides (TiO_2, Fe_2O_3, Cr_2O_3, SnO_2) mixed with 10 atom% Mo were also selective for the formation of acetone. Other metal oxides (V_2O_5, NiO, CuO, ZnO) were also found to yield acetone, but the selectivity was poorer. The presence of molyb-denum oxide appears to be essential for the reaction, WO_3 and U_3O_8 for example, when used as replacements for MoO_3, producing poor catalysts.

Of all the catalysts examined, the most effective have been based on tin-molybdenum oxides. Such catalysts are prepared [228] by mixing tin(II) hydroxide (precipitated from tin(II) chloride with OH$^-$) and ammonium molybdate to give a Sn/Mo ratio of 9/1, drying the resulting mixture and then heating it in air for two hours at 300°C. At temperatures between 115 and 135°C and in the presence of such catalysts, propylene is converted (conversion 3–9 %) to acetone with a selectivity of about 90 %. This catalyst can also be used to convert but-1-ene and but-2-ene selectively to methyl ethyl ketone, isobutene to tert-butanol and pent-2-ene to a mixture of methyl isopropyl ketone and diethyl ketone.

A. Mechanism

Buiten [229], following his initial discovery, attempted to elucidate the mechanism of acetone formation by investigating hydrogen-deuterium exchange between propylene and D_2O at 360°C in the presence of MoO_3/SnO_2 catalysts [429]. It was found that, in the presence of MoO_3, hydrogen-deuterium exchange was low: SnO_2 increased the rate of exchange, but in the presence of a SnO_2–MoO_3 mixture, the exchange was extremely rapid. Table 18 summarizes Buiten's results for the rate of this exchange reaction:

Table 18. Reaction rate constants for H–D exchange between propylene and D_2O

Catalyst	Surface area $(m^2 \, g^{-1})$	1st order reaction rate const.* $(mmole/h. \, m^2 \, atm.)$
MoO_3	9	5
SnO_2	21	55
SnO_2–MoO_3	4·6	2500

Apparently, one of the hydrogen atoms in propylene participated only slightly in the exchange and it was thought that this was the hydrogen atom located on the central carbon atom of the molecule. Infra-red measurements also revealed that deuterium appeared preferentially in the CH_2-group of propylene and mainly in a position cis- to the methyl group. To account for the exchange, a reversible reaction between propylene and an acidic, surface OH-group was proposed, giving a surface-bonded isopropoxy group:

* Buiten's choice of title here is very misleading. The quantity tabulated is not a rate constant and certainly not a first-order rate constant within the generally accepted definition. What has been recorded is the initial exchange rate divided by the propylene concentration.

The isopropoxy groups were considered to be the intermediates in the oxidation of propylene to acetone. The possibility that exchange coincides with the reaction:

$$H_2C\!\!=\!\!CH\!\!-\!\!CH_3 + H_2O \; \rightleftharpoons \; H_3C\!\!-\!\!\underset{\underset{OH}{|}}{C}H\!\!-\!\!CH_3$$

was discounted, although isopropanol was easily oxidized to acetone under the conditions adopted. Recently, Buiten [430] investigated the kinetic isotope effect in the oxidation of propylene over SnO_2–MoO_3 catalysts and found that propylene-2-d reacted 2·2 times less rapidly than ordinary propylene.

Moro-oka et al. have also attempted to elucidate the mechanism [228, 431, 432, 433] and experiments using H_2O^{18} proved that the oxygen atom in the product acetone comes, not from molecular oxygen, but from water. It was concluded that ketone formation proceeds via oxydehydrogenation of an alcohol or alcohol intermediate formed by olefin hydration. (Pralus et al. [434] have obtained a value of 12·2 kcal. mole^{-1} for the activation energy of propylene hydration to give isopropanol over Fe_2O_3–MoO_3 catalysts.) An attempt has also been made to define the active sites on Co–Mo and Sn–Mo catalysts. In the case of Co_3O_4–MoO_3 mixtures, although a compound $CoMoO_4$ has been found [228], it was thought not to be of significance. The active phase, it was reported, has an i.r. absorption band at ca. 725 cm^{-1}, and seems to involve an acidic point formed by the combination of, say, Co_3O_4 or SnO_2 with MoO_3. It should, perhaps, be reported here, that the solid-state reactions occurring in the Co_3O_4–MoO_3 [435] and Cr_2O_3–MoO_3 [436] systems (both effective catalysts for acetone formation) have been investigated. The reaction:

$$Co_3O_4(s) + 3MoO_3(s) \rightarrow 3CoMoO_4(s) + \tfrac{1}{2}O_2(g)$$

was reported to have a true rate constant of 5×10^{-3} s^{-1}, which was independent of composition (1:9 Co:Mo 9:1), and an Arrhenius activation energy of 53 ± 6 kcal. mole^{-1}. The rate-determining step was, apparently, diffusion of molybdenum ions through the product layer growing on the surface of the Co_3O_4 grains. In the reaction between Cr_2O_3 and MoO_3, a compound $(Cr_2Mo_3O_{12})$ was formed independently of the composition or sintering temperature of samples. Defect-type incorporation reactions were important in samples of low MoO_3-content ($<5\%$). In an oxidizing atmosphere, the excess oxygen content and p-type semiconductance were increased as a result of these reactions. ESR investigation showed that for samples with low Cr_2O_3-content, $Cr^{5+}H^+Mo^{6+}$ defects were formed in low concentrations ($<10^{-3}\%$).

The probable mechanism for the formation of acetone from propylene, it was concluded, can be summarized by the reactions:

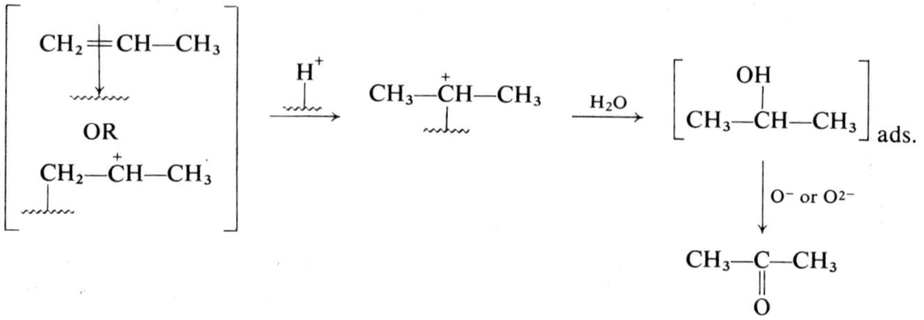

$$
\begin{bmatrix}
CH_2\!\!=\!\!CH\!-\!CH_3 \\
\text{OR} \\
CH_2\!-\!\overset{+}{C}H\!-\!CH_3
\end{bmatrix}
\xrightarrow{H^+}
CH_3\!-\!\overset{+}{C}H\!-\!CH_3
\xrightarrow{H_2O}
\begin{bmatrix}
OH \\
CH_3\!-\!CH\!-\!CH_3
\end{bmatrix}_{ads.}
$$

$$\downarrow O^- \text{ or } O^{2-}$$

$$CH_3\!-\!\underset{\underset{O}{\|}}{C}\!-\!CH_3$$

Finally it should be noted that Cant and Hall [157, 161] have oxidized propylene in the presence of certain noble metals and also detected significant amounts of both acetone and acrolein. Their findings are summarized in Table 19.

Table 19. Selectivity (%) towards partial oxidation products formed during propylene oxidation over silica-supported noble metals [157]

Metal	Acetone	Acrolein	Acetaldehyde	Acetic acid	C_3 acids
Ru	2–7	5–14	3–9	6–10	0·5
Rh	6–9	10–25	0·5–2·0	2–5	<0·5
Pd	2·5–4	1·0–3·0	<0·2	1·5–3·0	0·5
Ir	3–4	0·5–1·0	<0·2	28–32	2–3
Pt	0·2–0·7	0·1	<0·1	<0·4–0·9	<0·5

3.6. To Hexa-1,5-diene and Benzene

Under certain conditions, propylene undergoes oxidation to give either diolefins such as hexa-1,5-diene or aromatic compounds such as benzene, reaction to give the latter being known as dehydroaromatization. Reactions such as these have been shown to occur over a number of catalysts, including thallium(I) and thallium(III) oxides, indium(III) oxide [437], bismuth/tin oxides [438, 439] and Bi_2O_3–P_2O_5 [440] and modified manganese oxide catalysts [441]. Swift et al. [442] have also reported very high selectivities towards hexa-1,5-diene using Bi_2O_3 in the absence of oxygen; this reaction appears, however, to be non-catalytic, Bi_2O_3 acting as the oxidant in a typical gas–solid reaction.

At temperatures above 350°C, Trimm and Doerr [437] reported the selective oxidation of propylene to hexa-1,5-diene in highly fuel-rich mixtures. Carbon dioxide, a significant reaction product, was produced by further oxidation of the diene. It was concluded that thallium(III) oxide was the compound active in diene-formation and the following reaction scheme was postulated:

$$CH_3-CH=CH_2 \qquad CH_2=CH-CH_3 \qquad CH_2-CH_2-\overset{\ominus}{CH_2} \qquad CH_2-CH_2-\overset{\ominus}{CH_2}$$

$$\begin{array}{c} \xrightarrow{\hspace{1cm}} CH_2=CH-CH_2-CH_2-CH=CH_2 \\ + Tl^I \end{array}$$

followed by:

$$Tl^I + \tfrac{1}{2}O_2 \rightarrow Tl^{III} + O^{2-}$$

$$2OH^- \rightarrow H_2O + O^{2-}$$

Swift *et al.* [442] oxidized propylene/nitrogen mixtures using Bi_2O_3 as oxidant at temperatures between 475 and 520°C. The reaction was performed in cycles, the propylene mixture initially passing over the oxide for 10 minutes followed by air for a further 10 minutes to reoxidize the catalyst. Apart from hexa-1,5-diene, smaller amounts of hexa-1,3-, hexa-2,4-, and hexa-1,4-dienes, *cyclo* hexa-1,3- and -1,4-dienes were produced. Benzene was also formed, but only as a secondary product. At 475°C, it was found that the selectivity towards hexa-1,5-diene increased with increasing number of cycles over the oxide. Thus, in the first cycle, benzene and carbon dioxide (formed with selectivities of 35·2 and 64% respectively) were the main products. The second cycle produced hexa-1,5-diene with a selectivity of 26%, whilst the fourth and subsequent cycles produced this diene with a 41% selectivity. These changes were associated mainly with a loss of oxide surface-area, a process designated as "lining-out". At 520°C, a steady state selectivity of 76% towards hexa-1,5-diene was obtained, and the activation energy of the process was shown to be 27·5 kcal. mole^{-1}. In the presence of gaseous oxygen (i.e. with Bi_2O_3 functioning as a *catalyst*), Swift *et al.* [442] found that both the conversion and the selectivity for the process were lower. For example, propylene:oxygen:nitrogen (0·118: 0·047:0·835) mixtures formed a range of C_6 products (dienes and benzene) with an overall selectivity of 60%. It was concluded from these studies that the most probable mechanism for the reaction involved the initial formation of two allyl radicals, followed by their subsequent reactions in the gas phase.

In 1970, Sakamoto *et al.* [439] reported that significant yields of benzene were obtained during the oxidation of propylene ($O_2:C_3H_6 = 2:1$) at 500°C and in the presence of certain catalysts. The formation of benzene occurred with the phosphate, arsenate and antimonate of bismuth and also Bi_2O_3 + SnO_2. Apart from benzene, acrolein and CO_2 were produced, together with traces of carbon monoxide, acetaldehyde and hexa-1,5-diene. The best catalyst contained bismuth phosphate with a Bi/P ratio of 2:1, in the presence of which

40% propylene was converted to benzene. It was also noted that the oxidation of C_4 isomers (isobutene, but-1-ene, etc.) over this catalyst also gave aromatics, including benzene, ethylbenzene, toluene and the xylenes. The investigation of dehydroaromatization continued with the work of Seiyama et al. [440]. These workers investigated a number of metal phosphates (M = Ca^{2+}, Cr^{3+}, Fe^{3+}, Co^{2+}, Ni^{2+}, Cu^{2+}, Zn^{2+} and Ce^{3+}) and binary oxides containing bismuth or tin. It was found that the catalysts could be classified according to the reaction products formed; some gave acrolein and some yielded benzene. The eight metal phosphates were found to be without activity for the formation of benzene and, in the case of tin(IV) oxide, it was found that addition of basic oxides such as Na_2O resulted in catalysts effective for benzene formation; the addition of acidic oxides such as P_2O_5 favoured the formation of acrolein. In combination with Bi_2O_3, several oxides were tested, but only TiO_2, NiO, Sb_2O_4 and P_2O_5 produced catalysts yielding benzene. With P_2O_5, two combinations were very effective—the high temperature form of $BiPO_4$ (23% propylene converted to benzene) and a complex mixture described by the overall formula "$2Bi_2O_3 \cdot P_2O_5$" (35·4% propylene converted to benzene). The latter substance was found to consist of a very complex oxide mixture and contained γ-Bi_2O_3, $3Bi_2O_3 \cdot P_2O_5$, $2Bi_2O_3 \cdot P_2O_5$ and three forms of $BiPO_4$.

In the presence of oxygen, the dehydroaromatization reaction was considered to proceed via the steps:

(1) Oxidative dehydrogenation of propylene to give allylic intermediates.
(2) Dimerization of the intermediates to give diolefins such as hexa-1,5-diene.
(3) Aromatization of the dienes.

The catalytic activity was said to arise from the formation of the bismuth salt of an oxyacid. It was considered that Bi^{3+} combined with other metal ions through O^{2-} may act as adsorption sites for the allylic intermediates.

The mechanism oxydehydrodimerization has also been studied by Friedli et al. [441] and Dorogova and Kaliberdo [443], and Massoth and Scarpiello [444].

CHAPTER 4

The Catalytic Oxidation of C_4-Hydrocarbons

4.1. General Introduction

The C_4-fraction produced in the processing of petroleum is often a complex mixture of straight- and branched-chain alkanes and alkenes. A typical composition for such a fraction is shown below:

Component	W%
n-butane	7–12
isobutane	1–3
but-1-ene	26–28
but-2-ene	14–16
isobutene	45–47

From this material it is possible, by means of oxidation, to obtain commercially desirable products such as buta-1,3-diene, maleic acid, acetic acid and methacrolein. Several reviews have been compiled on the manufacturing processes involved and the uses of such compounds [446, 447] but the following chapter deals specifically with the catalysts involved.

4.2. Straight-Chain C_4-Hydrocarbons to Buta-1,3-diene

A major advance in the production of diolefins occurred when Hearne and Furman [448] showed that these compounds could be made from C_4- and higher alkenes by oxidative dehydrogenation in the presence of a bismuth molybdate catalyst. Since then, many different but effective catalysts have been discovered for this reaction and these function under a variety of conditions. For example, cobalt or nickel ferrite [449] act, not only as catalysts, but also as a source of oxygen; some catalysts (e.g. bismuth molybdate and bismuth tungstate) require the presence of steam in the feed-gases, whilst others (antimony–manganese oxide [450] and lead molybdate + cobalt or aluminium tungstate [451]) function in its absence. Some recent claims for suitable catalysts are shown in Table 20.

Industrial interest has stimulated numerous investigations into methods of improving the performance of dehydrogenation catalysts, particularly the influence of catalyst texture. Thus, Pitzer [479] noted the effect of steam-treatment on tin oxide–P_2O_5 (Sn:P = 1·5 molar ratio) catalysts and found that high-temperature steam had a beneficial effect on activity when such catalysts were used in the dehydrogenation of but-2-enes; increasing the steam-temperature from 1250 to 1600°F resulted in a three-fold increase in catalytic activity. A physical and chemical examination of steamed catalysts revealed that the increased activity was a result of a change in the macroporosity. Voge and Morgan [480] similarly examined the effect of the particle size of a Shell 205 catalyst (62·5% Fe_2O_3, 2·2% Cr_2O_3, 35·5% K_2CO_3) on the selectivity of the dehydrogenation, at 620°C, of but-2-ene in the presence of steam. Some dependence on particle size was observed and it was concluded that the process studied provided an example of the diffusional control of selectivity. Boutry *et al.* [481] also examined the influence of the degree of crystallinity on the activity of bismuth molybdate catalysts and found that, as the crystallinity decreased, so the specific activity of the catalyst increased. Finally, Shih [482] examined the effect on the dehydrogenation of butene of the variation of surface properties of silica, alumina-supported bismuth molybdate and silica-supported bismuth tungstate. Calcining silica-supported catalysts at 400°C and calcining the alumina-based catalyst at 800°C resulted in a process selectivity towards butadiene of 87%. It was found that the activity of silica-supported catalysts increased as the surface area and total pore-volume increased, whilst the opposite effect was observed with alumina-supported catalysts.

The production of diolefins by the oxidative dehydrogenation of alkanes is not generally successful, due to the low reactivity of these compounds. However, in the presence of halogens and their compounds, the dehydrogenation of butane to buta-1,3-diene has been satisfactorily carried out. Thus Polataiko *et al.* [483] have reported buta-1,3-diene yields of 58–78% in the dehydrogenation of both butane and the butenes in beds of mixed calcium oxide and iodide at 560–580°C, whilst Emel'yanova and Chugunnikova [484] have reported butadiene yields of 32–40 mole% at conversions of 50–55%, during butane dehydrogenation at 550°C in the presence of iodine (added as hydrogen iodide, iodine or ammonium iodide). Yields in the presence of other halogens were considerably lower. Skarchenko *et al.* [485] have also studied the oxidative dehydrogenation of butane in the presence of oxygen and added hydrogen iodide, bromine and iodine. At temperatures between 530 and 560°C, the addition of 0·05–0·06 mole I_2/mole butane increased the yield of butadiene relative to cracking products. Results indicated that of the three compounds, hydrogen iodide was the most efficient and hydrogen bromide was the least.

The efficiency of hydrogen iodide was thought to be due to the formation of iodine atoms by the reaction of the iodide with gaseous oxygen. Somewhat earlier, doubts had been raised by Skarchenko et al. [486] as to whether such reactions are actually heterogeneous and not exclusively gas-phase. In certain experiments, a butane, buta-1,3-diene, oxygen, nitrogen (1:0·03:1:10) mixture was passed through an empty reactor at 530–540°C and products containing 30 % butadiene were obtained. Coating the reactor with Al_2O_3 or NaCl lowered this figure to 20–25 %, whilst increasing the surface/volume ratio of the reactor (by packing with quartz chips) produced a concurrent increase in butadiene yield. Interesting results were also obtained by packing the reactor with silica gel (S = 30 m^2 g^{-1}) impregnated with 2 % of various metal oxides and halides. The large variations in butadiene-yield with different compounds, which one might expect for exclusively heterogeneous reactions were certainly not apparent; thus for the various compounds, the following butadiene yields (in brackets) were obtained: ZnO (63 %), MoO_3 (65 %), KNO_3 (63 %), Ag_2O (60 %) 2 % KOH + 2 % KI (62 %), 2 % Mn_3O_4 + 2 % MnI_2 (63 %), 1 % AgI + 5 % KBr (69 %), KI (75 %), BaI_2 (73 %). Finally, Pasternak and Badekar [487] have obtained theoretical evidence which suggests that sulphur or hydrogen sulphide–halogen mixtures would be equally effective substitutes for the much-more expensive iodine in dehydrogenation systems.

Kinetic and mechanistic studies performed on catalysts having high activity and selectivity will now be discussed.

A. Bismuth Oxide–Molybdenum Oxide Catalysts

The oxidation of butenes over bismuth molybdate catalysts has been the subject of considerable study. Over this catalyst, oxidation proceeds with a very high selectivity towards buta-1,3-diene, even at very high butene conversions; for example, Adams [488] reported selectivities of 90–95 % at conversions from 20–80 % in the oxidation of but-1-ene at 460°C. Numerous other products are also formed during the dehydrogenation, including but-2-ene (cis- and trans-), furan, maleic anhydride, acrolein, acetaldehyde, formaldehyde, acetone, methyl ethyl ketone, methyl isopropyl ketone, malonic, acetic and formic acids, and the oxides of carbon [184, 488, 489]. The but-2-enes are less reactive than but-1-ene [184, 490]; indeed, Roginskii et al. [491] have performed experiments in which mixtures of but-1-ene and ^{14}C-labelled cis- and trans-but-2-ene were passed over bismuth molybdate and concluded that but-1-ene is the primary source of butadiene. Table 21 summarizes the measured activation energies for the formation of some of the products during butene oxidation.

Over a range of binary catalyst compositions ($Bi_2O_3 \cdot 3MoO_3$, $Bi_2O_3 \cdot 2MoO_3$ (Erman phase), $Bi_2O_3 - MoO_3$) and ternary catalysts based on bismuth molybdate (e.g. $Bi_2O_3 - MoO_3 - Al_2O_3$), the oxidative dehydrogenation of the butenes has been reported to be first order in alkene and zero-order with respect to

Table 20. Some commercial catalysts suitable for the oxidative dehydrogenation of butene/butane mixtures to buta-1,3-diene

Authors	Catalyst	T°C (t_c secs)	But-1-ene	But-2-enes	i-butene	Butane	H₂O	O₂	N₂	C(%)	Product	Y(%)	S(%)	Ref.
Dow Chemical Co.	CaNi phosphate + 2% Cr_2O_3	500 (—)	74·7	+3·2 dibromobutane or Br_2	83·5	—	3000	100	—	38·7	Butadiene	—	92	[452]
Japanese Geon Co. Ltd.	Mo + Bi + Te + Sb (or Co) 12:1:1: 2:1 (P)	415 (—)	158·2	5 mole % comprising 29·3 15·8	—	—	3000	100	—	69·5	Butadiene / Methacr.	71 / 80	98	[453]
	Mo + Bi + Te + V + P 12:1:1:2	415 (—)	Composition as in Ref. [453]		46·1	8·8	—	7 mole %	88 mole %	—	Butadiene / Methacr. ac. 3 / Methacr. / Methacr. ac. 3	74 / 3 / 72 / 3	—	[454a]
	Mo + Bi + Te + P 10:2:1:1	420 (—)	Composition as in Ref. [453]							—	Butadiene / Methacr.	57 / 78	—	[454b]
Petro-Tex Chem. Corp.	97·5 % CeO_2 + 2·5 % LiCl	650 (—)	1 mole "butene" (+ 0·115 mol. Cl as HCl)		—		15 mol.	0·85 mol.	—	78	Butadiene	59	76	[455]
Inst. Franc. du Petrol	10% ZnO, 5% Fe_2O_3/ε -Al_2O_3	580 (0·2)		20 v %				10 v %	70 v %	24	Butadiene	76	—	[456]
		580 (1·0)		20 v %				10 v %	70 v %	26	Butadiene	60	—	
		580 (0·5)		21 v %				10 v %	69 v %	19	Butadiene	80	—	
		600 (0·2)		16 v %				20 v %	65 v %	40	Butadiene	72	—	
		630 (0·2)		10 v %				20 v %	70 v %	65	Butadiene	50	—	
		630 (0·5)		20 v %				20 v %	60 v %	54	Butadiene	53	—	
Boutry, Montarnal and Wrzyszcz	Bi_2O_3–MoO_3*	450 (—)												[457]
Esso Research and Engineering Co.	$AlPO_4$	593 (—)	—	1	1		—	1	—	24	Butadiene	—	—	[458]
	$AlPO_4$ + 2·5 mole % Mo (as Mo_4O_{11})	593 (—)	—	1	1		—	1	—	24	Butadiene	—	—	
Petro-Tex Chem. Corp.	Fe_2O_3 + Al_2O_3	621 (—)	—	1			20	0·5	—	34	Butadiene	26	76	[459]
	Fe_2O_3 + Al_2O_3†	621 (—)	—	1			20	0·5	—	46	Butadiene	40	87	
Bajars, L.	98% TiO_2 + 2% KCl/R.R.	700 (—)	—	1[a]			—	0·85	—	81	Butadiene	72	89	[460]
	95% MoO_3 + 5% BaO/R.R.	700 (—)	—	1[a]			—	0·85	—	88	Butadiene	71	81	
	72% MoO_3 + 28% CaO	688 (—)	—	1			12·5	0·7	—	—	Butadiene	—	91	
	91% TiO_2 + 9% CaO	623 (—)	—	1			12·5	0·7	—	—	Butadiene	—	81	
Dow Chemical Co.	CaNi phosphate‡	650–660 (—)	1 vol. "butene"				10–20 v	—	—	20	Butadiene	—	79–84	[461]
Phillips Petroleum Co.	Li (0·8 %), P (4·3 %), Sn (69 %) oxides	538 (—)		200 v/v cat./h			2400	air (1000)	—	95	Butadiene	90	—	[462]

* Catalysts prepared by melting Bi_2O_3/MoO_3 mixtures at 920–950°C, rapidly cooling and annealing at 560°C, then cooling and grinding at room temp. are claimed to be 10–12 times more active than known catalysts of similar composition when used in the oxidation of but-1-ene to butadiene.

Firm	Catalyst	Temp. °C (contact time)	Feed / conditions	Ratio	Conv. %	Product	Yield	Ref.
Phillips Petroleum Co.	P + Sn + Bi + O	1000°F(—)	200 v/v cat./h Composition as above	—	94 (70)	Butadiene 87 (54)	—	[462a]
	P + Sn + Bi + O + 1% B	—	Composition as above	2200	99 (87)	Butadiene 88 (64)	—	
	P + Sn + Bi + O + 3% B	—	Composition as above	air (1000)	100 (93)	Butadiene 88 (70)	—	
Phillips Petroleum Co.	P$_2$O$_5$ + SnO$_{28}$ (cont'g 10% P)	538 (—)	—	—	—	Butadiene	76§	[463]
Nippon Kagaku Co. Ltd.	Ni$_{2.5}$Co$_{4.5}$Fe$_3$BiP$_{3.5}$ K$_{0.03}$Mo$_{12}$O$_{54}$ on silica	305 (2·5)	1	5 / 10 (air)	—	Butadiene 88	52	[464]
Dow Chemical Co.	β-SrNi phosphate	400–700	200 v/v cat./h "butene" mixture + 50·3 v% O$_2$	—	50·9‖	Butadiene 42·5‖	83·4‖	[465]
	5–10% SrNi pyrophosphate + 90–95% β-Sr–Ni phosph.	640 (—)	21·3 steam/butene (vol. ratio)	—	46·0	Butadiene —	94·7	[466]
Kyushu Refractories Co. Ltd.	β-MgO:Al$_2$O$_3$:Fe$_2$O$_3$:CuO 40:32:8:18:2:4·5	650 (—)	800 v/v cat./h but-1-ene	—	63·2	Butadiene 50·6	80·0	[467]
Phillips Petroleum	* 2 catalyst beds employed, 1st contains Sn oxide-phosphate catalyst, 2nd contains same catalyst + 3% Li. Temp. also increases 850–1050°F	—	—	—	57·0	Butadiene —	96·0	[468]
Co.	27% Sn on an Al phosphate carrier	415 (—)	—	8	17·0	Butadiene —	99·0	[469]
Japanese Geon Co. Ltd.	Mo–Bi–Te–Sb–P 12:1:1:2:1 at.	457 (3·6) rat.	Hydrocarbon mixtures containing butenes and i-butene	6	—	Butadiene 70·9 / Methac. 79·5	83·8 / 84·3	[470]
	Mo–Bi–Te–Ag 1:1·6:0·3:0·1 at. rat.	—	1	2	78·2	Butadiene —	65·0	[471]
Ameripol Inc.	CaNi phosphate (98%) (Ca:Ni= 8·5) + 2% Cr$_2$O$_3$	700 (—)	Addition of small amounts of H$_2$S or n-butyl mercaptan, decreased hydrogenolysis and carbon formation and increased catalyst life	8	—	Butadiene	—	[472]
Petro-Tex Chem. Corp.	B-modified Mg ferrite	550 (—)	1·0	20 / 0·6	67·0	Butadiene —	92·0	[473]
	S-promoted Ba ferrite	—	—	—	—	Butadiene 50	—	[474]
	Fe$_2$O$_3$:MgO:CeO$_2$ 50:40:10	425 (—)	Present	—	65·0	Butadiene 62	95·0	[475]
	Mg ferrite + 2·5% P + 0·5% Ni	363–510 (—)	1·0	20 / 0·5	66·0	Butadiene 62·3	94·5	[476]
Gulf R. and D. Co.	M$_a$Cr$_b$Fe$_c$O where a = 0·1 to 3, b< 2 and c< 3; M = Mg, Ba, La, Ni, Zn, Cd	—	—	—	—	—	—	[477]
	MgCrFeO$_4$	325 (—)	—	—	—	Butadiene 69·0	91·0	[478]

Methacr. = methacrolein; Methac. = methacrolein; Methac. ac. = Methacrylic acid; ^{a} = 0·28 mole Br added as HBr; R.R. = catalysts supported on Vycor Raschig rings.

† = catalyst reactivated by the addition of silicone oil.
‡ = carbonaceous deposits build up on the catalyst.
§ = catalysts treated with steam at 677 to 843°C.
‖ In the absence of oxygen, C = 26·6%; S = 96·0% and Y = 25·6%.

Table 21. Activation energies (kcal. mole^{-1}) for the formation of certain products during the oxidation of n-butenes over bismuth molybdate catalysts

Product							
Buta-1,3-diene	Butene isomers	Maleic anh.	acrolein	CO	CO$_2$	CH$_3$CHO	Reference
26[a]	—	—	—	—	—	—	Adams et al. [184]
13[a]; 15[c]	4·5[a]	24[a]	25[a]	—	35	27[a]	Serebryakov et al. [489]
13[a, b]	—	—	—	—	—	—	Batist et al. [206]
16[d]; 27[e]	—	—	—	—	—	—	Sadovnikov et al. [492]
36[e]	—	—	—	—	—	—	Batist et al. [493]
10·5[d, f]; 33[e, i, g]	—	—	—	—	—	—	Keizer et al. [490]
10·0[d, g]; 15[e, j, g]	—	—	—	—	—	—	Ref. [490]
—	17–19[e]; 2·4[d]	—	—	—	—	—	Batist et al. [494]
—	12	—	—	—	—	—	Roginskii et al. [495]
—	—	—	—	—	24·8[g, k] 28·2[f, k] 28·1[h, k]	—	Matsuura and Schuit [211]
11,0[a, d, g]; 23·0[a, l]	—	—	—	—	—	—	Watanabe et al. [496]
10–14[a, g]; 34–44[a, e]	—	—	—	—	—	—	Watanabe et al., [497]

a = oxidation of but-1-ene; b = Fe$_2$O$_3$-promoted catalyst; c = oxidation of trans-but-2-ene; d = pulse-flow apparatus (uninhibited); e = recirculation apparatus (diene inhibited); f = catalysts with Bi : Mo = 1 : 1; g = catalysts with Bi : Mo = 2 : 1; h = catalysts with Bi : Mo = 2 : 3; i = T < 370°C; j = T > 410°C; k = oxidation of buta-1,3-diene; l = inhibited by isoprene.

oxygen over a wide range of temperature [489, 184, 490, 498]. Keizer [490], however, reported a slight deviation from zero-order with respect to oxygen with the low temperature koechlinite-modification of the catalyst, particularly at temperatures in excess of 385°C. In certain circumstances, deviation from first-order kinetics with respect to butene is also observed [493]. With both continuous- and recirculation-flow apparatus, Batist et al. [493] found that, at 348°C and with flow rates of less than 5 cm³ min⁻¹, the reaction was zero-order in butene; at flow-rates above this value, a first-order dependence was found. At 391°C, the more usual first- and zero-order dependence in butene and oxygen was observed. This behaviour was presumed to be linked with inhibition of the oxidation by buta-1,3-diene, a phenomenon which will be discussed later.

Finally, on the basis of conversion-selectivity data, Adams et al. [184] devised the following scheme for the oxidative dehydrogenation of but-1-ene:

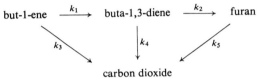

Values for k_3/k_1 of 0·05 and for $(k_2 + k_4)/k_1$ of 0·05 fitted the data well. A similar scheme has also been suggested by Tsailingol'd et al. [498], although it was suggested that the oxidation of both the mono-ene and the diene to carbon dioxide proceeded via intermediate, but unspecified, carbonyl compounds and acids (possibly acrolein, acetaldehyde, maleic, acetic or formic acids).

(a) Mechanism. Adams [488] postulated a mechanism for the oxidation of but-1-ene to butadiene over bismuth molybdate catalysts, which involved the initial adsorption of oxygen on the catalyst surface before reaction:

$$O_2(g) \longrightarrow 2O(ads.)$$
$$H_2C=CH-CH_2-CH_3 + O(ads.) \longrightarrow H_2C\text{···}CH\text{···}CH-CH_3 + OH(ads.)$$
$$H_2C\text{···}CH\text{···}CH-CH_3 + O(ads.) \longrightarrow H_2C=CH-CH=CH_2 + OH(ads.)$$

It now appears that this scheme is an oversimplification and, following the initial proposal by Sachtler and de Boer [499], that the oxidation is a classical example of the Mars–van Krevelen mechanism [500]. Thus according to Batist et al. [208], the catalytic oxidation of but-1-ene to buta-1,3-diene over both bismuth molybdate and molybdenum(VI) oxide, proceeds by the following mechanism:

$$C_4H_8 + \square + O^{2-} \rightleftharpoons C_4H_7^- + OH^- \left.\right\} \quad (1)$$
$$Mo^{6+} + C_4H_7^- \rightarrow [Mo=C_4H_7]^{5+} \left/\right.$$
$$[Mo = C_4H_7]^{5+} + O^{2-} \rightarrow Mo^{4+} + \square + OH^- + C_4H_6 \quad (2)$$
$$2OH^- \rightleftharpoons O^{2-} + \square + H_2O \quad (3)$$
$$O_2 + 2\square + 2Mo^{4+} \rightarrow 2O^{2-} + 2Mo^{6+} \quad \text{(fast)} \quad (4)$$

In reaction (1) and the subsequent steps, $\square$ denotes an anion vacancy and

$[Mo = C_4H_7]^{5+}$ represents a π-allyl complex of Mo^{6+} and the allyl carbanion $(C_4H_7^-)$. The kinetics of the reaction were reported to be first-order in but-1-ene and zero-order in oxygen, leading to the conclusion that reduction (steps (1) and (2)) is rate-determining and that reoxidation (step (4)), is comparatively fast. It is of interest that the reverse of reaction (1) leads to *cis*- and *trans*-but-2-ene i.e. double-bond isomerization although Alkhazov *et al.* [501] and Watanabe and Echigoya [502] believe that butene-isomerization involves a carbonium ion. The dependence of the catalyst composition on the activity and selectivity of the reaction was also investigated, and it was suggested that the activity was determined mainly by the Bi/Mo ratio (a maximum being attained with Bi/Mo = 1), whilst the selectivity was mainly determined by the occurrence of double-bond isomerization.

If this scheme is indeed correct, then the rate of catalyst reduction by but-1-ene should (i) proceed at a rate comparable with that of the oxidative dehydrogenation and (ii) lead to products observed under conditions of oxidation. This was investigated by Batist *et al.* [290] by recirculating but-1-ene/helium mixtures over certain catalysts (Bi_2O_3, MoO_3, non-stoichiometric molybdenum oxides and bismuth molybdate). At 530°C but-1-ene reduced Bi_2O_3 to metallic bismuth and although the reaction products consisted mainly of carbon dioxide, a small amount of butadiene and traces of the but-2-enes were also detected. At 538°C molybdenum(VI) oxide was reduced to molybdenum(IV) oxide and the reduction-rate was greater than that for bismuth(III) + bismuth(0). Under these conditions, the but-2-enes were the most important reaction products, although butadiene and the carbon oxides were also found. The rate of reduction of bismuth molybdate was faster than that of either component and butadiene was the principal reaction product; small, but equal, quantities of *cis*- and *trans*-but-2-ene were also formed. In the case of bismuth molybdate, the initial rate of reduction of the catalyst was the same as the rate of butene oxidation in the presence of gaseous oxygen and the reaction products were also identical.

In 1968, Batist *et al.* [205] published the details of a method for preparing bismuth molybdate catalysts having very reproducible properties. Catalysts were prepared by boiling slurries of freshly-precipitated, thoroughly-washed $BiO(OH) \cdot H_2O$ and H_2MoO_4 for about 2 hours. During boiling, both the colour and the viscosity of the slurries changed, and, after being dried and calcined at 500°C, the compound was found to be more active, selective and reproducible than previously-prepared samples. The highest activity was observed with a catalyst of composition $Bi_2O_3 \cdot MoO_3$, calcined for 2h at 500°C, which had an X-ray diffraction pattern very similar to that of koechlinite. Another excellent catalyst, of composition $Bi_2O_3 \cdot 2MoO_3$ and calcined at 600°C for 1h, had an X-ray pattern similar to that of a compound prepared somewhat earlier by Erman [503, 504]. Catalysts having a Bi/Mo atomic ratio of 2:3 and calcined for 2h at 500°C, were, however, much less active.

When catalysts containing low-temperature koechlinite were used in a kinetic investigation of butene oxidation, a remarkable change in the activation energy was observed as the temperature was decreased. Above 400°C, the activation energy was about 10 kcal. mole^{-1}, whilst below 400°C, it increased to about 30 kcal. mole^{-1}. This observation led Keizer et al. [490] to investigate thoroughly the pulse-reaction kinetics of but-1-ene oxidation over catalysts of various compositions, in the temperature range 260–425°C. The compound, $Bi_2O_3 \cdot 3MoO_3$, was found to have a higher activity towards but-1-ene isomerization than either the Erman phase of bismuth molybdate [503, 504] or low-temperature koechlinite; the general trend seemed to be that the more difficult the oxide was to reduce, the greater was its tendency to bring about isomerization of the double bond in the butenes. This, and later work [494] also showed that there was often a preference for the formation of cis-but-2-ene at low temperatures which shifted to a preference for trans-but-2-ene as the temperature increased. However, in all three cases, the activation energy for the oxidative dehydrogenation was 10–11 kcal. mole^{-1}.

Experiments were also performed in which butadiene was added to the gas-pulses. With the three compounds, it was found that butadiene retarded the reaction and increased the activation energy. This effect was most pronounced with the compound $Bi_2O_3 \cdot MoO_3$, for which the activation energy for the oxidation, at temperatures in excess of 410°C, was 15 kcal. mole^{-1} increasing to 33 kcal. mole^{-1} as the temperature fell to 370°C.

About the same time, Batist et al. [493] also reported a detailed study of the effect of buta-1,3-diene on the oxidation of but-1-ene over low-temperature koechlinite, using both conventional and recirculating flow apparatus. At 348°C, an inhibited reaction occurred with an activation energy of 37 kcal. mole^{-1}. However, at 391°C and above, diene inhibition was found to disappear and the activation energy fell to 11 kcal. mole^{-1}. Batist et al. [493] also performed interesting experiments on the reduction and reoxidation of the catalyst at 470°C. At this temperature, the yellow compound $Bi_2O_3 \cdot MoO_3$ was reduced, by but-1-ene, to a blue-black compound having the composition $Bi_2O_3 \cdot MoO_{2.5}$ (i.e. $Bi^{3+} + Mo^{5+}$). The X-ray diffraction pattern of the reduced material was very similar to that of the original catalyst, implying that reduction did not affect the position of the metal atoms. Even at temperatures greater than 470°C, the catalyst could not be reduced, using but-1-ene, beyond a state formally described by $Bi_2O_3 \cdot MoO_{2.5}$. Reduction by hydrogen gas, however, resulted in a compound containing bismuth metal and molybdenum(IV). It is interesting to note that catalysts produced by methods other than that described by Batist et al. [164], can also be reduced by butene gas to bismuth(0) and molybdenum(IV) [218], but no really satisfactory explanation of this behaviour can be offered.

Recently, Beres et al. [505] also investigated the reduction, in hydrogen at 360–500°C, of $Bi_2O_3 \cdot MoO_3$ and a series of bismuth molybdate catalysts.

Results obtained with α-$Bi_2O_3 \cdot 3MoO_3$, β-$Bi_2O_3 \cdot 2MoO_3$ and γ-$Bi_2O_3 \cdot MoO_3$ indicated a very rapid diffusion of oxygen through the crystal lattice. The activation energies for the reduction of the α-, β- and γ-forms of bismuth molybdate differed considerably, indicating large differences in the metal-O bond energies. Finally, Batist *et al.* [493] have concluded that the rate of reduction of koechlinite is equal to the overall rate of the reaction and strongly inhibited by buta-1,3-diene and that the rate of reoxidation of the reduced catalyst is fast and not inhibited by butadiene.

Very recently, Matsuura and Schuit [211] reported the results of studies of the adsorption equilibria of butenes, butadiene, water, etc. on some bismuth molybdates. These have led to a significant increase in our understanding of the mechanism of butene dehydrogenation. Both fully-oxidized and partially-reduced catalysts were investigated and these included $2Bi_2O_3 \cdot MoO_3$ (Bi/Mo = 4/1), the koechlinite modification $Bi_2O_3 \cdot MoO_3$ (Bi/Mo = 2/1), the Erman modification $Bi_2O_3 \cdot 2MoO_3$ (Bi/Mo = 1/1), $B_2O_3 \cdot 3MoO_3$ (Bi/Mo = 2/3) and MoO_3. On fully-oxidized catalysts, neither oxygen nor water was adsorbed; however, adsorption occurred with partially-reduced catalysts and, in contrast to the behaviour of the hydrocarbons, increased with the degree of catalyst-reduction.

Strikingly, the adsorption isotherms were found to follow simple Langmuir-Hinshelwood adsorption equations, either of a single-site (ss) or dual-site (ds) type:

(1) Single-site Langmuir-Hinshelwood adsorption isotherm (ss)

$$\frac{1}{v} = \frac{1}{v_m} + \frac{1}{v_m} \cdot \frac{p_0}{p}$$

(2) Dual-site Langmuir-Hinshelwood adsorption isotherm (ds)

$$\frac{1}{v} = \frac{1}{v_m} + \frac{1}{v_m} \cdot \frac{p_0^{1/2}}{p^{1/2}}$$

where

    v = amount of gas adsorbed

    v_m = amount of gas adsorbed at full surface-coverage

    p = equilibrium gas pressure at the end of the adsorption

and

    p_0 = gas pressure at half surface-coverage.

It was also possible to characterize adsorption in terms of the adsorbed volume (V_m) and the heat of adsorption (Q), according to the equation:

$$p_0 = p_0^0 \exp(-Q/RT)$$

From their results Matsuura and Schuit [211] concluded that, on bismuth molybdate catalysts, there exist two different surface sites for the adsorption of hydrocarbons and other gases. These sites were termed A-sites and B-sites. A-sites were single sites which were able to adsorb butadiene slowly but strongly ($V_m = 0.024$ cm^3 g^{-1}, $Q = 19$ kcal. mole^{-1}). Reduction of the catalyst produced a proportional increase in V_m and the A-site vacancies thus produced were able to adsorb some oxygen irreversibly and, more importantly, were able to adsorb water weakly and reversibly. It was concluded that an A-site is an O^{2-}-ion in a special position and is associated with a bismuth ion (MoO$_3$ contains no A-sites and Bi$_2$O$_3$ contains modified A-sites). B-sites were also found to adsorb butadiene but in a fast, reversible process representing weak adsorption. Similar behaviour was also observed with but-1-ene, propylene and cis-but-2-ene. These adsorptions were found to obey the dual-site equation, showing that a B-site must involve two adsorption centres. The value for V_m (0.1 cm^3 g^{-1}) in the case of the B-sites also suggested that there are considerably more B-sites than A-sites. It was thought that a B-site is comprised of an anion vacancy on a Mo^{6+}-ion surrounded by corner-shared O^{2-}-ions.

According to Matsuura and Schuit [211, 404], the active site for butene dehydrogenation contains one A-site and two B-sites. Since a B-site consists of two oxygen ions and an anion vacancy, the configuration of the oxygen ions in the site may be given as:

$O_B(1)$	$O_B(2)$	MoO$_3$	layer
	O_A	Bi$_2$O$_3$	layer
$O_B(3)$	$O_B(4)$	MoO$_3$	layer

The formation of butadiene from butenes may now be presumed to occur by the following steps:

In step 1, the adsorption of butene leads to the formation of a hydrogen atom and an allyl group (H-atom located on $O_B(1)$ and allyl group on $O_B(2)$). It was then postulated that step 2 involves the dissociation of a second hydrogen atom and the formation of a butadiene molecule weakly attached to $O_B(2)$ and $O_B(3)$:

At this point there are two alternative steps—either the weakly adsorbed butadiene is desorbed and the two hydrogen atoms ($H_{(1)}$ and $H_{(4)}$) migrate to O_A and are desorbed as a water molecule, or the weakly adsorbed butadiene becomes strongly attached at O_A and undergoes further oxidation:

$$(\text{Furan}) \longrightarrow CO, CO_2$$

These steps result in catalyst reduction. The catalyst is reoxidized at an A-site either by diffusion of an O^{2-} from the lattice or by gas phase oxygen.

(b) *Miscellaneous investigations.* A number of studies have been made of the effect of promoters on bismuth molybdate catalysts. For example, Morita *et al.* [508] added P, As, Sb, Cr, U and Fe to the catalysts and observed enhanced performance. The order of promoting ability was found to be:

$$Cr > Fe > P > U > As > Sb$$

Batist *et al.* [206] have also investigated the effect of P, Fe and Cr promoters, on the performance of the moderately active compound, $Bi_2O_3 \cdot 3MoO_3$. The effect of adding 10% mole ratio of $BiPO_4$, Fe_2O_3 and Cr_2O_3 to the catalyst was very pronounced and, in the case of $BiPO_4$ and Fe_2O_3, was due to the formation of koechlinite within the system. The Bi_2O_3–MoO_3–Fe_2O_3 system, in particular, was studied in detail and this system was found to contain four phases: $Bi_2O_3 \cdot 3MoO_3$, $Fe_2O_3 \cdot 3MoO_3$, koechlinite, and a new, but unidentified phase having a scheelite structure. The latter compound was found to be both active and selective in the oxidative dehydrogenation of but-1-ene and existed in the composition range $Bi/Fe = 8:2$ to $1:9$.

Russian workers have similarly studied the Mo–Bi–Fe system in some detail [506, 507]. Thus Annenkova *et al.* [507] studied both the physicochemical and catalytic properties of the system and, in particular, catalysts of composition $Mo:Bi:Fe = 6:2:2$ and $4:2:2$. Catalysts of composition $Mo:Bi:Fe = 4:2:2$, although more active in the oxidation of butene to butadiene than some unpromoted Bi-Mo catalysts ($Bi:Mo = 2:3$, $1:3$ and $1:1$), were also more active in the oxidation to carbon dioxide. An examination of the catalysts by i.r., X-ray diffraction and thermogravimetric methods revealed that the ternary system initially contained ferric molybdate, Bi_2O_3 and MoO_3. When the catalysts were heated, the following reactions were said to occur:

$$Fe_2O_3 + 3MoO_3 \rightarrow Fe_2(MoO_4)_3$$

$T \simeq 200°C \qquad Fe_2O_3 + Bi_2O_3 \rightarrow 2BiFeO_3 \text{ (or } Bi_2Fe_4O_9)$

$T > 450°C \qquad Bi_2O_3 + MoO_3 \rightarrow Bi_2O_3 \cdot 3MoO_3$

$T > 600°C \qquad Bi_2O_3 + Fe_2(MoO_4)_3 \rightarrow Bi_2O_3 \cdot 2MoO_3 + BiFeO_3 \text{ (or } Bi_2Fe_4O_9)$

It was therefore concluded that Bi–Mo–Fe catalysts were heterogeneous systems containing two or more chemical compounds in a ratio governed by the overall chemical composition of the catalyst and the conditions of preparation (firing temperature, time, etc.). For example, the compositions of the ternary systems studied were as follows:

Mo	:	Bi	:	Fe	
4	:	2	:	2	$Bi_2O_3 \cdot 3MoO_3$, $Fe_2(MoO_4)_3$, $BiFeO_3$ (or $Bi_2Fe_4O_9$)
					amorphous
6	:	2	:	2	$Bi_2O_3 \cdot 3MoO_3$, $Fe_2(MoO_4)_3$, $BiFeO_3$ (or $Bi_2Fe_4O_9$), MoO_3
					amorphous

Some work has been carried out on the effect of gaseous additives, particularly carbonyls, on the oxidative dehydrogenation of butenes over bismuth molybdate. Carbonyl compounds are usually co-products of the dehydrogenation reaction [184, 488, 489] and Adzhamov et al. [509] examined the effect of acetaldehyde, acetone, acrolein, methyl ethyl ketone and methyl vinyl ketone. Although saturated carbonyls had little effect, unsaturated carbonyl compounds inhibited both the isomerization and hydrogenation of butenes. A later report by the same authors [510] ascribed the enhancement of the rate of dehydrogenation on $Bi_2O_3 \cdot 4MoO_3$ at 425–525°C by gaseous ammonia, to removal of unsaturated carbonyls (as nitriles) by their reaction with ammonia. Oddly, Mal'yan et al. [511] have reported that acrolein (an unsaturated carbonyl) had no influence on the dehydrogenation reaction.

Finally, the oxidative dehydrogenation of butenes in the presence of bismuth molybdate has been studied by mathematical modelling [512, 513]. Thus Basner et al. [512] made a computer study of the effect of granule radius (from 0·1 to 0·5 cm) and temperature (from 430 to 530°C) on the reaction in the presence of Bi–Mo–P catalysts. It was predicted that an increase in catalyst granule-size leads to a decrease in the maximum obtainable yield of butadiene, particularly as the temperature increased.

B. Antimony Oxide-Containing Catalysts

Many catalysts containing antimony oxide are selective in the oxidation of butenes to buta-1,3-diene. Bakshi et al. [514] examined the catalytic properties of many Sb_2O_4–M_mO_n systems (M = V, Fe, Cr, Sn, Ni, Mn, Co, Cu) and found that the optimum composition lay within the limits $1:16 < M:Sb < 1:4$. They found that the most active catalyst contained V + Sb, the activity thereafter declining in the order: Fe + Sb > Cr + Sb > Sn + Sb $\doteq$ Ni + Sb > Mn + Sb > Co + Sb > Cu + Sb. The most selective catalyst was the system Fe + Sb oxides; the selectivity decreased in the order Fe + Sb > Mn + Sb > Sn + Sb > Co + Sb > Cu + Sb > Ni + Sb > Cr + Sb > V + Sb. The activity of antimony oxide-based catalyst seems to be highly dependent on surface area [515] and recently, Tarasova et al. [516] have reported the results of studies on the effect of method of preparation on surface area and phase composition of catalysts based on antimony oxide.

Of the antimony oxide-containing systems, iron(III) oxide–antimony oxide has been extensively studied, particularly by Boreskov and his co-workers [517–522]. This system was shown [517] to contain the free oxides of iron and antimony (α-Fe_2O_3, Sb_2O_4) as well as iron antimonate ($FeSbO_4$). Very recently, Malakov and Abdikova [523], by selective dissolution of the components, have obtained quantitative data on the concentrations of Fe_2O_3 and iron antimonate in such catalysts. These results showed that the maximum amount of antimonate was found in catalysts containing approximately 5·5 at. % Sb. A slightly earlier study, by Boreskov et al. [519, 522], into the effect of composition of Fe_2O_3–Sb_2O_4 catalysts on activity and selectivity, had shown that, in the composition range 3·2–42 % Fe_2O_3, the activity of the catalysts changed little with composition and the selectivity of the process towards butadiene was at a maximum. This behaviour was ascribed to the formation of iron antimonate.

Boreskov et al. [518, 521] have studied both the pulse oxidation of but-1-ene over Fe–Sb–O catalysts at 425°C, and the oxidation-reduction of the catalyst surfaces under these conditions. A fully-oxidized catalyst was found to be very unselective in the production of butadiene (selectivity towards diene = 10 % and towards carbon dioxide = 70 %), but, as the catalyst became reduced, its activity decreased but the selectivity towards butadiene increased to 90 % and thereafter remained steady, It was concluded that the reaction did not proceed with the alternating reduction and oxidation of the original catalyst, but involved the reduction and oxidation of an already partially-reduced surface. It was calculated that the catalysts performed with maximum selectivity towards butadiene, when almost 40 % of an oxygen monolayer had been removed from the surface. It was also concluded that, in the oxidation of butene, two types of oxygen participate. One type (strongly bound to the catalyst) was thought to yield the diene, whilst the other (weakly bound) was thought to yield carbon monoxide and dioxide. Boreskov et al. [522] also obtained kinetic data on Sb oxide–Fe$_2$O$_3$ (>42 %) catalysts, and found that reaction orders, in the presence of these catalysts, were 0·5 in both oxygen and butenes.

The pulse-oxidation method was also employed by Ammonsov *et al.* [524] to study the formation of carbon monoxide and carbon dioxide during the oxidative dehydrogenation of but-1-ene. ^{14}C-labelled but-1-ene and buta-1,3-diene were oxidized in the presence of various Fe–Sb–O catalysts and the following overall reaction scheme was deduced

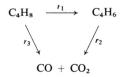

Their kinetic data are summarized in Table 22. Activation energies for the formation of carbon dioxide were also derived, and these are given below

Catalyst	E_A (kcal. mole^{-1})
α–Fe$_2$O$_3$	23
α–Fe$_2$O$_3$, FeSbO$_4$	21·6
FeSbO$_4$	20
Sb$_2$O$_4$	—

Finally, Shchukin and Ven'yaminov [525] have assessed the contribution of a homogeneous component to the oxidation of butenes over Fe–Sb–O catalysts.

Catalysts based on the antimony oxide–tin(IV) oxide system also have been fairly well investigated [515, 526–530]. Sekushova *et al.* [515] studied the activity, in the oxidative dehydrogenation, isomerization and deep oxidation of C$_4$/C$_5$ olefin mixtures, of thirteen Sn–Sb–O catalysts, ranging in composition from pure tin(IV) oxide to pure Sb$_2$O$_4$. In the temperature range 350–450°C, tin(IV) oxide and Sb$_2$O$_4$ had little activity in the reaction and the maximum activity was shown by catalysts within the composition range 4:1–9:1 Sn/Sb atomic ratio. The activity of the catalysts was proportional to their specific surface areas, and the active component of the catalysts was amorphous under X-ray examination. Trimm and Gabbay [527] investigated the kinetics and mechanism of the oxidation of the butenes over a catalyst of composition 1:4 Sn/Sb. At 440°C, butadiene was the main product, although some isomerization and complete oxidation also occurred. Other products observed in small quantities included acrolein, acetaldehyde and butyraldehyde. Traces of acetylene, ethylene, propylene, methacrolein, isobutyraldehyde, methyl ethyl ketone, furan and benzoic acid were also detected. The following reaction

Table 22. The oxidation of but-1-ene (1-C_4 : O_2 : N_2 = 5 : 10 : 85 v%) and buta-1,3-diene over Fe–Sb–O catalysts.

Catalyst	Composition (mole% SbO)	Temp. (°C)	†Formn. of CO via		†Formn. of CO_2 via	
			r_2	r_3	r_2	r_3
α-Fe_2O_3	0	365–427	$9 \cdot 10 \times 10^{-1}$ to $1 \cdot 50$	$8 \cdot 50 \times 10^{-1}$ to $3 \cdot 40$	$10 \cdot 9$ to $23 \cdot 6$	$36 \cdot 1$ to 112
$FeSbO_4$, α-Fe_2O_3	28	365–453	$5 \cdot 30 \times 10^{-1}$ to $1 \cdot 62$	$1 \cdot 34 \times 10^{-1}$ to $4 \cdot 05$	$6 \cdot 07$ to $18 \cdot 2$	$13 \cdot 4$ to $77 \cdot 0$
$FeSbO_4$	66	350–475	$1 \cdot 16 \times 10^{-2}$ to $5 \cdot 6 \times 10^{-1}$	$1 \cdot 35 \times 10^{-2}$ to $1 \cdot 86 \times 10^{-2}$	$6 \cdot 05 \times 10^{-2}$ to $1 \cdot 48$	$1 \cdot 16 \times 10^{-1}$ to $8 \cdot 9 \times 10^{-1}$
Sb_2O_4	100	445–451	$5 \cdot 7 \times 10^{-5}$ to $4 \cdot 7 \times 10^{-6}$	$1 \cdot 5 \times 10^{-5}$ to $1 \cdot 6 \times 10^{-5}$	$1 \cdot 1 \times 10^{-5}$ to $8 \cdot 8 \times 10^{-5}$	$1 \cdot 2 \times 10^{-5}$

† r_2 and r_3 in mmole min^{-1} m^{-2}

r_2 and r_3 of the form:
$$r_3^{CO_2} = k_3[1\text{-}C_4H_8][O_2]^{1/2}$$
$$r_2^{CO_2} = k_2[C_4H_6][O_2]^{1/2}$$

network was devised to explain the observed product spectrum:

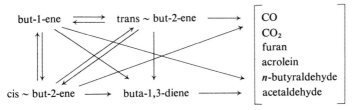

Trifiro and Pasquon [528, 529] similarly studied the oxidation of butene over Sn–Sb–O catalysts; however, although at temperatures within the range 300–460°C dehydrogenation of but-1-ene occurred, isomerization did not. A dependence of the reaction rate on the oxygen partial pressure was also noted. If this is the case, such catalysts are very different in their behaviour to Bi–Mo–O catalysts. Alieva et al. [530] investigated the effects of the elements Li, Na, K and Rb on the catalytic activity of Sn–Sb–O (Sn:Sb = 4:1) catalysts. Unlike the other metals, small quantities of lithium promoted the oxidative dehydrogenation of but-1-ene. With respect to the isomerization of but-1-ene to but-2-enes, the concentration of alkali metals decreased the reaction monotonically, suggesting the neutralization of Lewis acid-type sites. According to their effect on the degree of isomerization, the metals tested were listed in the order:

$$Li > Na > K > Rb$$

The oxidative dehydrogenation of butenes over uranium oxide–antimony oxide catalysts has recently been studied by Nishikawa et al. [531] who prepared the catalysts of various U:Sb atomic ratio, by precipitating ammonium uranate onto antimony oxide and calcining the resulting mixture in air at 500–1000°C for 4–12 hours. In the presence of such catalysts, the oxidative dehydrogenation of but-1-ene at 450°C was examined. The activity of the catalysts was influenced, not only by the U:Sb atomic ratio, but also by the temperature of catalyst-calcination. For example, samples calcined at a high temperature or samples having a low U:Sb atomic ratio were most selective towards the formation of butadiene. An X-ray examination of samples calcined at the higher temperatures revealed the formation of some unspecified U–Sb compound.

Simons et al. [247] prepared uranium oxide–antimony oxide catalysts by two different methods. In the first method (I), the catalyst was precipitated (with NH_4OH) from a solution prepared by adding liquid antimony(V) chloride to concentrated hydrochloric acid, followed by uranyl acetate. The second method (II) involved a similar precipitation from a solution prepared by the addition of uranyl acetate to a solution of Sb_2O_3 dissolved in hydrochloric acid. The initially-formed precipitates varied in colour, depending on the Sb:U ratio, from orange to yellow–brown. After being boiled with water, dried and sintered for 16 hours at 725°C, catalysts were obtained, varying in colour from dark green (excess U), through blue–violet (Sb:U = 1 to 2), to yellow-brown (excess Sb). Catalysts having Sb:U ratios of 0·5, 1·0, 2·0, up to 10, were prepared in this way and their behaviour in the oxidation of but-1-ene depended both on the Sb:U ratio and the method of preparation. (It is interesting to note that catalysts with a Sb:U ratio less than 4 are unstable when prepared by method (I).) As in a similar investigation by Callahan and Grasselli [249], Simons *et al.* [247] reported the identification of the compounds $(UO_2)Sb_3O_7$ and $U_3Sb_3O_4$ and believed the former to be the active component. (Later work by Grasselli *et al.* [250] revealed that this compound is better described by the formula (USb_3O_{10}). A previously unreported compound—Sb_6O_{13}—was also identified [247].

At first sight, the oxidation of but-1-ene over U–Sb–O catalysts is similar to its oxidation over Bi–Mo–O, i.e. the reaction is first-order in hydrocarbon and zero-order in oxygen. At temperatures in excess of 350°C, an activation energy for the reaction of 22 kcal. mole^{-1} was derived from continuous-flow measurements, but pulse-kinetic studies yielded a value of 8 kcal. mole^{-1} [247]. With certain catalysts (Sb:U = 4) and at temperatures below 350°C, there was also evidence of inhibition by a reaction product, although the inhibitor was not buta-1,3-diene in this case. Great behavioural differences do apparently exist, however. Thus, bismuth molybdate can undergo extensive reaction with butenes in the absence of oxygen and may then be completely restored by reoxidation [218]. Uranium–antimony oxide, on the other hand, reacts only slightly with butenes in the absence of oxygen but even this leads to structural changes in the catalyst which cannot be reversed by gaseous oxygen [247]. One very fundamental difference in the behaviour of the two catalysts is the absence of double-bond isomerization in the presence of U–Sb–O catalysts. The implications of this are great: if isomerization is connected with the formation of an allylic intermediate in the dehydrogenation reaction, then either the reaction in this particular case does not involve an allylic species or, the formation of allylic intermediates does not always lead to isomerization as previously assumed.

C. Mixed Metal Oxides with a Spinel Structure

Since the publication of a patent specification by Kehl and Rennard [532] for the use of ferrite-based catalysts in the oxidation of butenes to buta-1,3-diene, the results of a number of studies on their use as dehydrogenation catalysts have been published in the non-patent literature [533–535]. Cares and Hightower [533] reported studies on the use of the inverse spinels $CoFe_2O_4$ and $CuFe_2O_4$ in dehydrogenation, whilst Rennard and Kehl [534] and Massoth and Scarpiello [535] all investigated zinc chromium ferrites and magnesium chromium ferrites of different composition. Some of the catalysts are quite active and selective as Table 23 indicates.

Table 23. Catalytic properties of some spinels in the oxidation of but-2-ene to butadiene at 325°C [534]

Catalyst	Conversion (%)	Selectivity (%)
$ZnCrFeO_4$	58	91
$ZnCr_{0.25}Fe_{1.75}O_4$	56	90
$ZnCr_{0.1}Fe_{1.9}O_4$	46	92
$ZnFe_2O_4$	20	89
$ZnCr_2O_4$	15	16
$MgCrFeO_4$	64	90
$MgFe_2O_4$	53	86
$MgCr_2O_4$	28	32
α-Fe_2O_3	35	83

The kinetics and mechanism of dehydrogenation in the presence of spinel catalysts are quite complex and both surface and lattice oxygen atoms appear to participate in the reaction. For example, cobalt and copper ferrites are ineffective in the isomerization of butenes and the order of the dehydrogenation reaction is less than one in both butene and oxygen, and is strongly inhibited by buta-1,3-diene. In the case of cobalt ferrite, however, the selectivity of the oxidation is strongly dependent on the amount of gaseous oxygen present, in contrast with the behaviour observed with copper ferrite [533].

In order to gain further insight into the mechanism of dehydrogenation, particularly in the presence of zinc chromium ferrite, Massoth and Scarpiello [535] studied the ease of reduction of this and related oxides using hydrogen and but-1-ene. In the presence of hydrogen, the reducibility of the oxides decreased in the order:

$$Fe_2O_3 > ZnFe_2O_4 > ZnCrFeO_4 = FeCrO_3 > ZnCr_2O_4 > ZnO > Cr_2O_3$$

Whilst in the presence of but-1-ene, the following reactions occurred:

$$Fe_2O_3 \rightarrow Fe_3O_4 \quad (T = 400–540°C)$$
$$FeCrO_3 \rightarrow Fe_3O_4, Cr_2O_3 \quad (T = 540°C)$$
$$ZnCr_2O_4 \rightarrow \text{no reaction} \quad (T = 400°C)$$
$$\rightarrow C \text{ deposition} \quad (T = 500°C)$$
$$ZnFe_2O_4 \rightarrow \text{Surface-O loss} \quad (T = 400°C)$$
$$ZnCrFeO_4 \rightarrow \text{Surface-O loss} \quad (T = 400°C)$$
$$\rightarrow C \text{ deposition} \quad (T = 500°C)$$

There seems to be little doubt that oxidative dehydrogenation over metal oxides involves a redox cycle and the utility of a catalyst depends on its ability to undergo reduction without lattice-collapse, so that oxidation restores the original state. It is noteworthy, therefore, that the two oxides which showed only surface reduction ($ZnFe_2O_4$, $ZnCrFeO_4$) also exhibit excellent selectivity towards butadiene formation [534]. In the case of zinc chromium ferrite,

Massoth and Scarpiello [535] confirmed a redox cycle involving Fe(II) and Fe(III) by means of e.s.r. measurements on the reduced catalyst and also suggested that the presence of both zinc and chromium stabilizes the iron towards reduction beyond iron(II). The reaction mechanism proposed by Massoth and Scarpiello [535] was similar to that proposed by Rennard and Kehl [534]:

$$C_4H_8 \quad \begin{array}{c} C_4H_8 \\ \square \\ O \diagdown \diagup O^- \\ Fe^{3+} \end{array} \longrightarrow \begin{array}{c} C_4H_7 \quad H \\ O \diagdown \downarrow \diagup O \\ Fe^{3+} \end{array} \longrightarrow \begin{array}{c} H \quad C_4H_6^- \quad H \\ O \diagdown \diagup O \\ Fe^{3+} \end{array} \longrightarrow C_4H_6$$

$$\tfrac{1}{2}O_2 \quad \begin{array}{c} \square \\ O \diagdown \diagup O \\ \square \\ Fe^{2+} \end{array} \longleftarrow \begin{array}{c} \square \\ O \diagdown \diagup H_2O \\ \square \\ Fe^{2+} \end{array} \longleftarrow \begin{array}{c} H \quad \square \quad H \\ O \diagdown \diagup O \\ C_4H_6 \\ Fe^{2+} \end{array} \quad H_2O \quad C_4H_6$$

The active centre for the adsorption of but-1-ene was postulated to a combination of an anion vacancy ($\square$) and a sorbed oxygen radical ion (O^-). Hydrogen-abstraction from the adsorbed butene then gave the species C_4H_7 and OH^- and a further hydrogen-abstraction yielded butadiene and another OH^-. During this step iron(III) was reduced to iron(II) and the two hydroxyl ions combined to give water. The iron(II) was then reoxidized, either by gas-phase oxygen or by adsorbed but uncharged oxygen. Although this sequence is similar to that postulated by Schuit [218] for bismuth molybdate catalysts, the difference between the two mechanisms is the reaction involving O^- rather than lattice O^{2-} ions.

D. Other Mixed Oxide Catalysts

An investigation of the dehydrogenation of straight-chain alkanes in the presence of a chromia-alumina catalyst, suggested to Pines and Goetschel [536] that ring intermediates on the surface of the catalyst were involved in the reaction. It is interesting, therefore, to find other evidence suggesting that the formation of butadiene from butenes might also involve ring intermediates [537, 538]. Okamoto et al. [537] found that the principal products from the dehydrogenation of $1\text{-}^{14}C$-but-1-ene over acidic chromia-alumina catalysts were n-butane, cis- and trans-but-2-ene, buta-1,3-diene and unchanged $1\text{-}^{14}C$-but-1-ene. At 485°C, the yield of butadiene was 9·5%, increasing to 69·3% at 675°C. The distribution of radioactivity in the products showed that they were formed via an intra- rather than an inter-molecular mechanism and, particularly in the case of butadiene, that rearrangement of the labelled atom had occurred (from the 1-position in but-1-ene to the 2-position in butadiene). The extent of the rearrangement was determined and, at 485°C, was about 6%, rising to 10% at 550°C and 23% at 675°C. Several possible mechanisms were considered in an attempt to explain the re-arrangement, including: (i) recombination of C$_2$-fragments produced by butene pyrolysis; (ii) the formation of a 3-ring intermediate; and (iii) the formation of a *cyclo*butane- or *cyclo*butene-type intermediate. Mechanism (i) was quickly discounted on account of the small extent of pyrolysis, and mechanism (ii) was rejected since it predicted the formation of isobutene (which was not detected):

$$
\overset{*}{\underset{1}{C}}=\underset{2}{C}-\underset{3}{C}-\underset{4}{C}
\quad\xrightarrow{\ -H\ }\quad
\left[\ \underset{1}{\overset{*}{C}}\underset{\substack{\diagup \ \diagdown \\ (+) \\ C_3 \\ | \\ C_4}}{}\ C_2\ \right]
$$

$$
\overset{*}{\underset{1}{C}}-\underset{2}{C}=\underset{3}{C}-\underset{4}{C}
$$

$$
\begin{array}{l}
\xrightarrow[\ +H\ ]{C_1 + C_3} \quad \overset{*}{C}=C-C-C \\[4pt]
\xrightarrow[\ +H\ ]{C_2 + C_3} \quad C=\overset{*}{C}-C-C \\[4pt]
\xrightarrow[\ +H\ ]{C_1 + C_2} \quad \overset{*}{C}\diagdown_{C}\diagdown_{C}=C
\end{array}
$$

Mechanism (iii) involved the formation of four-membered ring-intermediates. Thus, dehydrogenation of butenes at the 1- or 4-position would make a *cyclo*butane-type intermediate on the catalyst surface. Cleavage of either the 2–3 or the 3–4 bond would then produce rearrangement of the carbon-14.

$$
\overset{*}{\underset{1}{C}}=\underset{2}{C}-\underset{3}{C}-\underset{4}{C}
\quad\xrightarrow{\ -H\ }\quad
\left[\ \begin{array}{c} C-C \\ \| \quad | \\ {}^{*}C \quad C+ \end{array}\ \right]
\quad\longrightarrow\quad
\left[\ \begin{array}{c} \overset{2}{C}+\overset{3}{C} \\ |\ (+)\ | \\ {}^{*}\underset{1}{C}-\underset{4}{C} \end{array}\ \right]
\quad\xrightarrow{\ +H\ }\quad
\begin{array}{l} C=\overset{*}{C}-C-C \\ \text{or} \\ C-C=C-\underset{*}{C} \end{array}
$$

$$
\overset{*}{\underset{1}{C}}-\underset{2}{C}=\underset{3}{C}-\underset{4}{C}
$$

An alternative suggestion was that the buta-1,3-diene initially formed was further dehydrogenated in the 1- or 4-position forming a *cyclo*butene-type intermediate. Rupture of the 2–3 bond in such an intermediate would produce the required rearrangement:

$$
\overset{*}{C}=C-C=C
\quad\xrightarrow{\ -H\ }\quad
\left[\ \begin{array}{c} C-C \\ \| \quad | \\ {}^{*}C \quad C+ \end{array}\ \right]
\quad\longrightarrow\quad
\left[\ \begin{array}{c} C+C \\ \| (+) | \\ {}^{*}C-C \end{array}\ \right]
\quad\xrightarrow{\ +H\ }\quad
C=\overset{*}{C}-C=C
$$

Delgrange and Blanchard [538] found that the oxidation of butene (3% in air) at 400°C over a V$_2$O$_5$–25% MoO$_3$ catalyst yielded sixteen products at butene conversions of less than 20%. Of these products, the C$_4$-fraction contained buta-1,3-diene, maleic acid, isobutyraldehyde, methacrolein, methyl ethyl ketone, methyl vinyl ketone, crotonaldehyde, furan, and buta-2,3-dione. Oxidation of H-^{14}C-but-1-ene yielded similar products, but part of the initial labelled atom was found on the central carbon atom of methacrolein and part on the central carbon atoms of maleic acid. Delgrange and Blanchard explained these results and the formation of isobutyraldehyde, in terms of an initial, partial rearrangement of but-1-ene or but-2-ene to give methyl-*cyclo*propane.

In conclusion, Russian workers have investigated certain aspects of catalysis involving CaNiCr phosphate, a catalyst developed by the Dow Chemical Co. [452, 461]. For example, Andrushkevich *et al.* [539] studied the kinetics of the oxidative dehydrogenation of butene over these catalysts, whilst Afanas'ev and Buyanov [540] studied the regeneration of coked catalysts which had been employed in butene-dehydrogenation at temperatures in the region of 570–650°C. In the latter study, catalyst regeneration was carried out using either steam-Ar-oxygen or Ar-oxygen mixtures and it was found to be a very complex process, depending on temperature, catalyst properties (surface area, etc.), distribution of the carbon over the catalyst and the composition of the regenerating gas.

4.3. N-Butenes

The oxidation of butenes to carbonyl compounds is an interesting reaction, which has been known for some time. Thus Clark and Shutt [541] disclosed that metal selenites and tellurites could be employed in the oxidation of but-1-ene to methyl vinyl ketone (but-1-en-3-one) and the oxidation of but-2-ene to crotonaldehyde. Similarly, Hearne and Adams [542] showed that copper(I) oxide was an active catalyst in the oxidation of both but-1-ene and but-2-ene to methyl vinyl ketone.

The oxidation of butenes over copper(I) oxide, over a range of temperatures has been studied by Popova and Mil'man [543], by Gorokhovatskii [544], and by Radzhabli *et al.* [545, 546]. The oxidation of the butene isomers was found to proceed at a similar rate in all cases. Compared with the oxidation of propylene or isobutene on similar catalysts, the oxidation of but-1-ene and the but-2-enes proceeds with a higher activation energy (18 kcal. mole^{-1}, cf. 12–13 kcal. mole^{-1}) and yields a more diverse mixture of reaction products. Amongst these methyl vinyl ketone (50–63% formed) is often obtained in good yield and crotonaldehyde (8–16%) to a lesser extent. A number of other products can also be detected, including buta-1,3-diene, acrolein, 3-ethyl oxiran, 2,3-dimethyl oxiran, small amounts of saturated carbonyl compounds (acetaldehyde, acetone, propionaldehyde, methyl ethyl ketone) and the unsaturated alcohol, methyl vinyl carbinol. Apparently, the total yield of carbonyl compounds is independent of the position of the double bond [545] although Radzhabli *et al.* [546] have also reported that the yield of methyl vinyl ketone from but-1-ene is higher. Finally, the reaction yielding carbonyl compounds was found [546] to be independent of butene concentration but first-order in oxygen (cf. the reaction yielding buta-1,3-diene which has an order of 0·5 in butene).

In the low-temperature oxidation of but-1-ene in the presence of 9:1 SnO$_2$–MoO$_3$ and 9:1 Co$_3$O$_4$–MoO$_3$ catalysts, Tan *et al.* [228] observed the selective production of methyl ethyl ketone. For example, at 135°C and in the presence of a 9:1 SnO_2–MoO_3 catalyst, a 83·5% yield of methyl ethyl ketone was obtained from but-1-ene at a conversion of about 5%. As the temperature increased, the yields of methyl ethyl ketone declined, falling to 60·3% at 160 and 38·7% at 180°C. Over the same catalyst, *trans*-but-2-ene was less selectively oxidized, forming 60·6% methyl ethyl ketone at 130 and 32·8% at 200°C. Co_3O_4–MoO_3 catalysts were found to be slightly less active and selective, in the oxidation of but-1-ene, than the tin-containing compounds (60·8% methyl ethyl ketone at 240°C, declining to 27·8% at 274°C), and much less selective in the oxidation of the but-2-enes (*trans*-but-2-ene yielded 26·7% methyl ethyl ketone at 233°C and *cis*-but-2-ene gave 14·5% at 266°C).

4.4. Butane and Butenes to Maleic Anhydride

The oxidation of C_4-fractions of petroleum hydrocarbons to maleic anhydride is a commercially desirable process and a number of catalysts described in the patent-literature are given in Table 24.

Although there are a number of catalysts described in the patent-literature, reports of other, non-patent, studies are much fewer. A number of investigations have, however, been carried out by Ai *et al.* on the V_2O_5 and V_2O_5–P_2O_5 systems [555–558] and on the P_2O_5–MoO_3 system [559, 560]. A silica-supported V_2O_5–P_2O_5 system has also been recently studied by Akimova *et al.* [561] and Ostroushko *et al.* [562]. Ai *et al.* [555–557] examined the partial oxidation of *cis*-but-2-ene, buta-1,3-diene and furan over pure V_2O_5 and V_2O_5–P_2O_5 mixtures to ascertain the effect of the addition of P_2O_5 to V_2O_5 on the activity and selectivity of such catalysts in butene oxidation. The addition of P_2O_5 was found to lower the activity of the catalysts although the selectivity increased. Thus, in the presence of a catalyst of P/V ratio = 1·6, the selectivity, towards but-1,3-diene, of *cis*-but-2-ene oxidation increased from 43% (V_2O_5 alone) to 88%. Similarly, the selectivity of the oxidation of buta-1,3-diene to furan increased from 72 to 93 mole%, although the selectivity of the oxidation of furan to maleic acid fell from 60 to 50 mole% (a phenomenon ascribed to an increase in "polymer-forming reactions" [556]). On the basis of such results, Ai *et al.* have suggested the following oxidation mechanism:

Table 24. Some commercial catalyst configurations for the oxidation of butane and butenes to maleic anhydride

Authors	Catalyst composition	T(°C) (t_c sec.)	Feed-gas composition	C(%)	Y(%)	Ref.
Princeton Chem. Res. Inc.	P–V–O (P : V = 1·75)	400 (—)	"butene"—air	—	—	[547]
I.C.I. Ltd.	Mo–TeO_2	425 (1·5)	2% butenes (92% 2-C_4H_8), 50% H_2O, 10% O_2, 38% N_2	—	30	[548]
I.C.I. Ltd.	18% V_2O_5–P_2O_5 (4 : 5 molar ratio, supported on Al_2O_3 having a range of surface area e.g. S_A = 75 m² g⁻¹, 20 m² g⁻¹, 1·5 m² g⁻¹, 0·4 m² g⁻¹, <0·1 m² g⁻¹	272/322 (2·5), 248/275 (2·5), 385/423 (2·5), 500/524 (0·8), 450/500 (1·2)	0·7 v% butene in air, 0·7 v% butene in air, 1·0 v% butene in air, 0·6 v% butene in air, 0·7 v% butene in air	88, 70, 100, 100, 100	12·8, 21·6, 25·9, 36·4, 46·7	[549]
I.C.I. Ltd.	Cobalt molybdate/glass helices	400 (6)	1·5% c-2-C_4H_8, 35% H_2O, 63·5% air	—	47, 29^a	[550]
Mitsubishi Chem. Ind. Co. Ltd.	V_2O_5–P_2O_5 (1 : 2)/silica	405 to 420	4 mole% "butene" mixture in air (mixture = 18·4% 1-C_4H_8, 13·0% 2-C_4H_8, 31% C_4H_6, 7% C_4H_{10}, 29·0% iso-C_4H_8).		46	[551]
Petro-Tex Chem. Corp.	20% V : P : Nb† 1·0 : 1·3 : 0·0275 on alundum	406	0·7 to 0·75 mole% 2-C_4H_8 in air		88	[552a]
	V : P : Cu : Nb 1·0 : 1·35 : 0·082 : 0·02 on alundum	446	0·7 mole% 2-C_4H_8 in air		84	[552b]
	P–V–Cu–Nb–Li 1·43 : 1·0 : 0·036 : 0·022 : 0·068	446	As above		87	

Catalyst	Temp. (°C)	Conditions	Result	Ref.
V : P : Cu : Li 1·0 : 1·4 : 0·024 : 0·072		As above	92	[552c]
P : V : O + 0·003-0·125 atom K per atom of P	480		87	[552d]
20% V_2O_5 : P O (1 : 1·2) + 1% by wt. Li/alundum	550	0·7 mole% 2-C_4H_8 in air (92 g 2-C_4H_8/l cat./h)	93	[552e]
B.A.S.F. A.-G. 4% V_2O_5, 10% P_2O_5, 86% TiO_2	452	8·62 g/h (1l/h) "butene" mixt. (57% 1-C_4H_8, 7·6% c-2-C_4H_8, 12·8% t-2-C_4H_8)	7·7 bg/h	[553a]
2-5% V_2O_5, 5-10% P_2O_5 80-88% anatase	390		56·5	[553b]
B.A.S.F. A.-G. V_2O_5 (8·1%), WO_3 (2·5%) P_2O_5 (11·9%), NiO (4·5%), TiO_2 (73%)	450	29·3 g/h "butenes" + air (1200 l/h)	25·5g/h	[553c]

† Reactivation of such catalysts using organo-phosphorus compounds R_3P, $R_3P=O$; $(R(OX)_2P=O$, $R_2(OX)P=O$, $R(OP)(OX)_2$ etc. where R=C_6H_5 or C_1 to C_6 alkyl and X=R or S or R, is described in Ref. [554]. a = in absence of steam

In this reaction scheme, crotonaldehyde was a suggested intermediate in the formation of furan from butadiene and indeed, the participation of both crotonaldehyde and crotonic acid in the oxidation of butenes to maleic anhydride has received considerable support [563–565]. According to Ai [555, 557], the oxidation of furan is the principal factor determining the rate of formation of maleic acid and the selectivity of the process. In this context, Ai *et al.* [557] also studied the activity of V_2O_5-based catalysts towards the further oxidation of maleic anhydride. At 500°C on pure V_2O_5, 25% maleic anhydride underwent complete oxidation whilst on P_2O_5-containing catalysts, the decomposition was very much less extensive (2% with catalysts having $P/V = 1.6$). It was concluded, therefore, that the beneficial action of P_2O_5 is due to its control of side reactions occurring parallel to the formation of maleic anhydride.

Recently, Ai and Suzuki [566] reported the effect of the addition of Bi_2O_3 to MoO_3–P_2O_5 catalysts, on the partial air-oxidation of butene, butadiene and furan to maleic acid. Oxidation was carried out in the presence of catalysts of fixed P:Mo atomic ratio (0.2), but a varying Bi:Mo ratio (0, 0.05, 0.10, 0.15, 0.2, 0.3, 0.4, 2, 4, ∞). In the temperature range 350–450°C, catalytic activity in the oxidation of *cis*-but-2-ene, for example, was strongly affected by the addition of even a small quantity of Bi_2O_3. Thus, at 450°C, a MoO_3–Bi_2O_3–P_2O_5 (1:0.1:0.2) mixture is about four times more active than MoO_3–P_2O_5 alone. The activity reached a maximum, however, at $Bi/Mo = 0.1$ and declined at higher ratios. Similarly, in the oxidation of butene, butadiene and furan, the highest yields of maleic acid (30, 60 and 76 mole% respectively) were also obtained with catalysts having a Bi/Mo ratio equal to 0.1.

The V_2O_5–MoO_3 system has been extensively studied by Blanchard and Delgrange [538, 567–570] and also by Jescai [571]. In the oxidation of but-1-ene, at 405°C, over a series of V_2O_5–MoO_3 catalysts supported on non-porous α-Al_2O_3, Blanchard and Delgrange [567] observed that the maximum values for activity and selectivity towards maleic acid were obtained with catalysts containing 25 mole% MoO_3. Catalysts with this composition also had the highest activity in the isomerization reaction in the absence of air and at the same temperature. In a further account of this work [568], rates were reported for the conversion of both but-1-ene and a mixture of but-2-enes (64% *cis*- and 36% *trans*-) during oxidation. These values were 19.6 mole h^{-1} g cat.$^{-1}$ for but-1-ene and 17.8 mole h^{-1} g cat.$^{-1}$ for the but-2-enes. With both butenes, apart from oxidation to maleic acid, Blanchard and Delgrange [568] observed direct oxidation to carbon dioxide, oxidation to methyl ethyl ketone and, surprisingly, to methacrolein. Later reports [538, 569] on the oxidation of 4-^{14}C-but-1-ene over V_2O_5–30% MoO_3 catalysts, propose the formation of cyclic intermediates (methyl*cyclo*propane and *cyclo*butane) during the oxidation. Recently, Blanchard *et al.* [572] have studied the isotopic exchange of

$^{18}O_2$ on a series of V_2O_5–MoO_3 catalysts and have found that the activation energy of the exchange reaction varies with the composition of the catalyst. The activation energy was found to have its highest value for a catalyst containing 30% MoO_3, a composition very close to that required for the most selective oxidation of butenes.

The use of Co–Mo–oxide catalysts in the oxidation of butenes to maleic anhydride has been investigated by a number of workers, including Sadykhova et al. [573], Avietsov et al. [574] and Boutry et al. [575]. With such catalysts, Sadykhova et al. [573] found that, in the temperature range 380–440°C, the rate of formation of maleic anhydride was proportional to the butene concentration but independent of the oxygen concentration. The activation energy for the process was calculated to be 11·2 kcal. mole^{-1}, a value close to that obtained by Slovetskaya et al. [576] for the chemisorption of butane on the surface of a Cr_2O_3–Al_2O_3–K_2O dehydrogenation catalyst. Avietsov and co-workers [574] oxidized a C_4-mixture containing but-1-ene, trans-but-2-ene and buta-1,3-diene to maleic anhydride on a Co–Mo–O catalyst (Co : Mo = 1·7) at 350 and 400°C and found that, with oxygen concentrations in excess of 17 v%, the process parameters were independent of oxygen. At 350°C, the total conversion of but-1-ene greatly exceeded that of the but-2-ene or butadiene although, at 400°C, the values were very similar. At 400°C, the conversion of the C_4-hydrocarbons into oxygen-containing products was 64, 70 and 81% for but-1-ene, but-2-ene and buta-1,3-diene, respectively. Finally, Boutry et al. [575] have investigated Co–Mo–O catalysts for use in the dehydrogenation of butane. The catalytic activity depended on the Co : Mo ratio and the catalyst structure. Using X-ray diffraction techniques, the catalysts were found to contain monoclinic $CoMoO_4$, hexagonal cobalt molybdate and a γ-phase. A surface redox mechanism was also proposed for the reaction, and this involved oxygen from the γ-phase crystal lattice.

In conclusion, a number of investigations have been carried out by Agasiev et al. [577–579] into the direct conversion of butane into maleic anhydride. Initially, Agasiev et al. [577] examined both Co–Mo–O and V–Mo–O catalysts in the temperature range 420–500°C. The best results were obtained using a V–Mo–O catalyst, although the yield of maleic anhydride was quite low (18 w% based on the amount of butane reacted). Later, Agasiev et al. [578] developed a complex catalyst consisting of cerium chloride and Co–Mo oxides supported on silica. The function of the cerium chloride was to dehydrogenate the butane to butenes which were then converted to maleic anhydride by the Co–Mo oxides. In this way, a higher yield (29·8 mole% at 490°C) of maleic anhydride was obtained and, by the addition of 0·59 mole hydrogen chloride per mole of butane, Agasiev et al. [579] obtained a yield of anhydride of 62 w% over the same catalyst at 490°C. Finally, the kinetics of the oxidation of butane have been reported by Kuz'michev and Skarchenko [580].

Table 25. Some catalysts suitable for the oxidation of C$_4$-hydrocarbon fractions to acetic acid

	Catalyst composition	T(°C)	Feed-gas Composition	C(%)	Y(%)	Ref.
Krabetz	Mo + W + V oxides + (opt. Fe, Ti, Al, Cu, Mn, Ni, Sn)	250	1500 v/h, "butenes"; 40,000 v/h air; 25,000 v/h steam	95	87	[587]
Kuraray Co. Ltd.	V$_2$O$_5$ + 73% rutile	245	721/h C$_4$-fraction (0.6% C$_3$'s + 11.5% nC$_4$H$_{10}$ + 2.5% iC$_4$H$_{10}$ + 25.5% 1-C$_4$H$_8$ + 14% 2-C$_4$H$_8$ + 45% i-C$_4$H$_8$); 2970 l/h air; 1420 l/h steam	87	34.2	[588]
Japan Synth. Chem. Ind. Co. Ltd.	V–Ti–P–O 1 : 1.07 : 0.05 : 4.67	220 to 235	1 vol. C$_4$ + 32.3 v air + 16.3 v steam	—	49.8	[589]
	V–Cr–O 1 : 0.1 : 2 to 8	—		—	49.2	[590]
	V$_2$O$_5$–TiO$_2$ (V : Ti = 1 : 1)	—	C$_4$-fraction : air : steam = 1 : 28.9 : 16.3. C$_4$-fraction = 27.8% 1-C$_4$H$_8$; 46.9% i-C$_4$H$_8$; 6.8% C$_4$H$_{10}$; 15.7% 2-C$_4$H$_8$; 1.1% i-C$_4$H$_{10}$	—	0.617 moles /mole C$_4$	[591]
Montecatini Edison S.p.A.	Mo + V (+ Sb)	355	2-C$_4$H$_8$ (3.5%) + O$_2$ (11.2%) + N$_2$ (32.8%) + H$_2$O (52.5%)	—	50.0	[592]
	Mo + Sn (75 : 25)	210	2-C$_4$H$_8$ (1.1%) + O$_2$ (14.0%) + N$_2$ (34.9%) + H$_2$O (50.0%)	95.2	S = 55%	[593]
Chemische Werke Huels A.-G.	V–Ti	190 to 260	10 l/h C$_4$-fraction (53% 1-C$_4$H$_8$, 27% 2-C$_4$H$_8$, 19% C$_4$H$_{10}$); 500 l/h air; 280 l/h steam	80.0	72.5	[594]
	Ti vanadate	185 to 200	"Butene" (>, 0.8 v%) mixed with air or oxygen and steam	—	—	[595]
Toa Fuel Industry Co. Ltd.	V : W (1 : 0.2 at. rat.)	280	"Butene" : air : steam = 1 : 39 : 10	42.3	49.3	[596]

Stamicarbon N.N.	Sn–Mo–O	230	4%v "butenes" + 3 moles O_2 + 18·89 moles H_2O	43·0	34	[597]
	Sn–Mo–O	>200	"Butenes", 26 l/l cat./h + O_2, 213 l/l cat./h + N_2, 715 l/l cat./h + steam, 473 l/l cat./h	81·0	42	[598]
	SnO₂–MoO₃–SiO₂ 26 : 37 : 37%	280	2·25%/v "butene" + O_2 + steam	83·0	S = 43	[599]
Kuraray Co. Ltd.	V + Ti + Zn (V–Ti = 1 : 10–10 : 1); Zn–(V + Ti) = 0·01 : 1	250	—	—	33	[600]
	V + Sn + Zn	255	—	—	35	[601]

4.5. C_4-Hydrocarbons to Acetic Acid

The oxidation of butane and butenes to acetic acid seems to be a reaction in which industrial organizations, alone, have an interest. Apart from a report by Brockhaus [581], little information has appeared outside patent claims. Catalysts involved in this process consist of vanadates of metals such as Ti, Al [582], Sn [583], Sb [584, 585] and Zn [586] and, at temperatures in the region 240–270°C, fairly high yields of acetic acid may be obtained. For example, at 270°C and in the presence of a titanium vanadate catalyst (Ti:V = 1:0·98), a butene–air–steam (1:30:17) mixture yielded 70% acetic acid at a butene conversion of 73%. Other products were maleic acid (3 mole%) and the oxides of carbon (25 mole%) [582]. Table 25 lists other, very recent, patent claims.

4.6. Isobutene to Methacrolein and Methacrylonitrile

Isobutene behaves very similarly to propylene under conditions of catalytic vapour-phase oxidation. Just as acrolein (prop-1-en-3-al) and acrylonitrile are produced from propylene under the appropriate conditions, so methacrolein (2-methylprop-1-en-3-al) and methacrylonitrile can be obtained from isobutene, often by using the same catalyst under similar conditions (for example, Refs. 290, 619). Table 26 illustrates this point quite well in the case of acrolein and methacrolein.

Table 26. Conversion-selectivity data for the oxidation of propylene and isobutene to their respective aldehydes over a Bi–Mo–O catalyst at 460°C. [619]

Alkene	Conversion (%)	Selectivity (%)
Propylene	10	90
	40	80
	70	73
Isobutene	10	72
	40	72
	70	72

Significant differences in behaviour do exist, however. For example, Uchijima and Oda [602] oxidized propylene and isobutene to their corresponding unsaturated aldehydes in the presence of both bismuth molybdate and bismuth phosphomolybdate catalysts and found that, at low alkene conversions, both processes were first-order in alkene and independent of oxygen. Up to a point, this shows that the same mechanism is involved in aldehyde formation, although at hydrocarbon conversions in excess of 30%, considerable deviations from first-order kinetics were observed with isobutene. According to both Uchijima and Oda [602] and Adams [619], isobutene is more easily oxidized

Table 27. Some patented catalysts for the oxidation of i-butene to methacrolein

Authors	Catalyst composition	Carrier	T°C (t_c secs)	Feed-gas composition				C(%)	Y(%)	S(%)	Ref.
				$i\text{-}C_4H_8$	O_2	N_2	steam				
Mitsubishi Rayon Co. Ltd.	Mo:Co:Te:Ca	—	440 (1·5)	7·5%	62·5% air		30%	28·2	—	82·5	[603]
	Mo:Co:Te:Ca	—	460 (1·5)	As above				39·8	—	80·8	
	Mo:Co:Te:Na	—	460 (1·5)	As above				42·2	—	81·4	
	Mo:Co:Te	—	435 (1·5)	As above				38·5	—	35·6	
	Mo:Te:Na (or K):Fe (or Sb, Ca, Bi)	—	420 (3·0)	8·0%	62·0% air		30%	80·0	—	80·5	[604]
Kanegafuchi Spinning Co. Ltd.	Cu–Li	SiO_2–Al_2O_3 grains							"high"	—	[605]
Japan Oil Co. Ltd.	$AgVO_3$ or Ag_3VO_4 or $Ag_4V_2O_7$	Silicon carbide	400 (—)	60	40 air				19	—	[606]
Petro-Tex Chem. Corp.	V:P:B = 1:14:0:1	Alundum or SiC	490 (0·9)	1·1%	98·9% air				32	—	[607]
Toyo Soda Manufg. Co. Ltd.	V_2O_5:Bi_2O_3:$Mo(W)O_3$ 1·6 mole%:4·36 mole%:40·4 mole%	Fused Al_2O_3	—						20	—	[608]
Gulf Oil Corp.	$Cr_9PMo_{12}O_{52}$ or $Cr_9PV_{12}O_{46}$ or $Cr_9PW_{12}O_{52}$	—	381 (5·6)	11·1 v%	58·9 v% air		30 v%	12	20	—	[609]
		Porous SiC	442 (2·0)	13·6 v%	74·0 v% air		12·4 v%	25	57	—	[609a]
	7% $Cu_9S_2PMo_{12}O_{51.5}$		487 (2·3)	As above				43	63	—	
			534 (1·8)					44	53	—	
Petro-Tex Chem. Corp.	$Cu_9Te_{1.1}PV_{12}O_{43.7}$	Porous SiC ($<10\ m^2\,g^{-1}$)	538 (2·1)	14·4 v%	73·2 v% air		12·4 v%	42	46	—	[609b]
	Ag_2O (3 pts)+P_2O_5 (1–7 pts)	Alundum	500 (0·27)	10%	10%		80%		61	—	[610]
	Cu + P_2O_5	Alundum	600 (0·09)	10%	7·5%		82%		72†, 68‡, 47·8§	—	[610a]
B.F. Goodrich Co.	Pyrophosphates of Ag, Fe and/or Cu	Firebrick, kieselguhr, Al_2O_3	480 (10·7)	1 g mole	1·97 g mole		4·7 g mole	87	62	70	[611]
Rohm u. Haas G.m.b.H.	$VOSO_4$:TeO_2:MoO_3 1:0·6:1	—	450 (—)	1	14·3 air		9		38	49	[612]
Mitsubishi Rayon Co. Ltd.	MoO_3+Sb_2O_5 (+ Ca, Zn, Mg, Sr, Cd, or Ba)	—	470 (—)	6·7%	63·3% air		30%	84	65	—	[613]
	MoO_3+Fe_2O_3+0·1% TeO_2	—	380-450 (2·5)	7·6	15	57	20			—	[613a]
	Sb_2O_5+P_2O_5	—	450 (0·9)	6·8%	63·2% air		30%	91	30	—	[613b]
Oda et al.	Mo, Bi+silicic acid	—	450 (0·9)	1	9 air				35	—	[614]
Narita and Nishijima	Cu+Na+O	Celite	315	50	13	8	37			—	[615]
Japanese Geon Co. Ltd.	Mo+Bi+Fe+Ag+O		435 (3·6)	1	2		6	93	60	65	[616]
Nippon Kayaku Co. Ltd.	Ni:Co:Fe:Bi:P:K:Mo:O 2:5:4:5:3:0·1:0·5:0·07:12:54	—	345 (—)	1	2·2		5·0	98	70	—	[617]
Japan Synth. Chem. Ind. Co.	Hastelloy-B (28% Mo, 61% Ni, 5% Fe) flakes	—	540 (—)	1	23·5 (air)				68	—	[618]

† 0·5% H_2S added.
‡ 0·5% SO_2 added.
§ control (no additives).

than propylene (rate constant for methacrolein formation/rate constant for acrolein formation = 4·5–13·2 at 450°C [602], and, although the oxidation of propylene is markedly affected by a phosphorus promoter (apparent activation energy lowered from 23 kcal. mole^{-1} to 14 kcal. mole^{-1}), the oxidation of isobutene seems to be hardly affected (activation energy constant at 27·28 kcal. mole^{-1}) [602]. Probably on account of such differences, some catalysts are claimed to be effective only in the oxidation of isobutene to methacrolein [603] and Table 27 lists other such catalysts.

A. Kinetics and Mechanism

(a) Copper oxide catalysts. Since the discovery by Hearne and Adams [174] of the effectiveness of cuprous oxide as a catalyst in isobutene oxidation, a considerable number of kinetic and mechanistic investigations have been carried out. For example, Mann and Rouleau [620] studied the kinetics of the oxidation to methacrolein on a pumice-supported copper oxide catalyst at 350–450°C and found that the rate of isobutene oxidation could be expressed by the equation:

$$\text{Rate} = \frac{kK_1[C_4H_8][O_2]}{1 + K_1[C_4H_8] + K_2[C_4H_6O]}$$

It was further suggested that the rate-controlling reaction was that between adsorbed isobutene and gaseous, or weakly-adsorbed oxygen. Gorokhovatskii et al. [621, 622] have also studied this reaction, and at 320°C, found that the products included methacrolein, acrolein, acetaldehyde, propionaldehyde, carbon dioxide and water [621]; activation energies for methacrolein and carbon dioxide formation have been determined as 15 and 20 kcal. mole^{-1}, respectively [622].

It has long been known that the specificity of propylene oxidation, over copper catalysts, can be improved by the presence of gaseous additives such as halogens [2, 181] and Mann and Yao have made a number of studies to elucidate the effect of similar modifiers for the oxidation of isobutene over copper oxide. The compounds investigated included selenium dioxide [623], sulphur dioxide [624] and halogenated hydrocarbons [625, 626]—compounds having a higher electron affinity than the skeleton catalyst. The effect of several variables (including the modifier:copper ratio, the oxygen:isobutene ratio and the temperature) on the reaction was elucidated and the results are summarized in Table 28. From this table it can be seen that even very small amounts of modifiers improve the activity and selectivity of the catalyst, although there is an optimum concentration, beyond which the reverse is true.

As a result of a study of the oxidation of isobutene and methacrolein in the presence of both halogen-modified and unmodified catalysts, Mann and Yao [625, 626] concluded that, under normal reaction conditions, the major part of carbon dioxide produced in isobutene oxidation originated from the further

Table 28. Influence of modifiers on the conversion and selectivity towards methacrolein in the oxidation of isobutene over copper oxide.

Additive	Additive:Cu ratio	T(°C)	iC$_4$H$_8$: O$_2$: N$_2$			C(%)	S(%)
				Feed (moles/hr.)			
SO$_2$	0·0000	400	0·39 : 0·21 :	0·80		13·7	38·4
	0·0230	400	As above			16·0	48·1
	0·0000	375	As above			8·3	42·6
	0·0230	375	As above			10·8	53·5
	0·0420	375	As above			11·1	45·9
SeO$_2$	0·0000	425	0·39 : 0·28 :	—		21·1	33·8
	0·0067	425	As above			29·7	39·0
	0·0200	425	As above			29·0	44·8
CH$_3$Br	0·0000	400	0·36 : 0·21 :	0·80		14·8	40·2
	0·00009	400	As above			18·0	43·1
	0·00070	400	As above			15·6	78·8
	0·00070	400	0·36 : 0·30 :	1·1		20·4	77·2
	0·00070	400	0·36 : 0·34 :	1·3		22·1	80·4
C$_2$H$_2$Cl$_4$	0·00000	400	0·11 : 0·11 :	0·41		20·1	24·8
	0·00050	400	As above			34·1	67·3
	0·00100	400	As above			37·7	73·9
	0·0025	400	As above			37·9	80·3
C$_3$H$_7$I	0·00030	400	As above			34·4	70·7
	0·00050	400	As above			41·7	76·6
	0·00070	400	As above			37·4	77·2
	0·0030	400	As above			21·4	14·4

oxidation of methacrolein. In the case of halogen modifiers, their beneficial effect on selectivity was due to suppression of this undesired further oxidation. According to Mann and Yao [625] modifiers function by controlling the density of positive holes (pL) and free electrons (eL) on the catalyst surface. For example, when isobutene (a donor gas) is adsorbed on copper(I) oxide (a p-type semi-conductor), the concentration of charged, adsorbed isobutene is low on account of the scarcity of positive holes. Oxygen, on the other hand, being an acceptor gas is adsorbed in much greater concentrations due to the ready supply of electrons from the valence-band of the catalyst. Trace amounts of modifiers with electronegativities greater than copper act, therefore, by decreasing the electron supply, thereby increasing the number of positive holes and the catalyst selectivity.

(b) *Other mixed oxides.* Zhiznevskii and co-workers have extensively examined the catalytic properties of a numer of ternary oxide systems (and their binary sub-systems) with respect to the oxidation of isobutene to methacrolein. Such systems include Fe–Te–Mo oxides [627, 628], Fe–Sb–Mo oxides [628, 629] and a number of catalysts based upon a 4:1 Sb:Te oxide mixture [630–632]. In studies involving the Fe–Te–Mo system, Zhiznevskii *et al.* [627] examined the activity and selectivity of a number of catalysts within the composition range 21–62 atom % Fe, 20–50 atom % Mo and 0–46 atom % Te. Each component influenced the properties of the catalyst differently: for example, an increase in the molybdenum content up to about 22–35 atom % increased catalytic activity, whilst an increase in tellurium content up to 30 atom % improved catalytic selectivity. However, only if the iron content of the catalyst exceeded 35 atom %, was the catalyst mechanically strong enough to be useful. The best results were obtained using a catalyst of composition Fe:Mo:Te = 1:1:0·85 and a study of the kinetics of isobutene oxidation in the presence of steam and in the temperature range 360–420°C, enabled the following relationships to be derived:

$$\text{Rate of formation of methacrolein} = \frac{k_1[O_2][H_2O]^{0\cdot4}}{1 + b[C_4H_6O]} \qquad \text{moles } s^{-1} \, 1^{-1}$$

$$\text{Rate of formation of } CO_2 \quad = \frac{k_2[O_2][H_2O]^{0\cdot25}}{1 + b[C_4H_6O]}$$

$$\text{Rate of formation of CO} \quad = k_3[O_2][H_2O]^{0\cdot25}$$

k_1, k_2 and k_3 are unspecified constants.

Activation energies for the formation of methacrolein, carbon dioxide and carbon dioxide were found to have the values 32, 39 and 45 kcal. mole^{-1}, respectively.

The relationship between the catalytic properties of the Fe–Sb–Mo systems, sub-systems such as Sb–Mo, Fe–Sb, etc., and pure oxides (Fe_2O_3, Sb_2O_4, MoO_2) has been studied by methods including X-ray diffraction and EPR [629, 633]. With respect to the formation of methacrolein from isobutene, the following series of activities and selectivities were obtained:

	Activity	Selectivity
Single Oxides	$Fe_2O_3 \gg MoO_2 > Sb_2O_4$	$Sb_2O_4 > MoO_2 \gg Fe_2O_3$
Binary Oxides	$Fe\text{–}Mo \simeq Sb\text{–}Mo > Fe\text{–}Sb$	$Sb\text{–}Mo > Fe\text{–}Sb > Fe\text{–}Mo$

With the Sb–Mo system, addition of iron increased the catalytic activity; the amount required for maximum activity was 38·5 atom % but, for maximum selectivity, 24 atom %. Systems containing Fe–Sb oxides alone were found to have both a low specificity (to methacrolein) and overall activity, although addition of up to 38 atom % Mo increased both activity and selectivity; beyond this concentration, the overall activity towards complete oxidation increased, whilst the specific activity altered little. Optimum ternary catalysts were found to have the composition Fe:Mo:Sb = 1:1:0·6. Catalysts based on the Sb–Te (4:1) system and also single oxides have been studied by Zhiznevskii *et al.* [630–632]. Their kinetic data are shown in Tables 29 and 30.

Table 29. Specific rates of formation (W_{sp}) of oxidation products in the oxidation of isobutene at 400°C in the presence of various oxides [630]

| | $W_{sp} \times 10^7$ mole m^{-2} s^{-1} | | | | | |
Catalyst	Sb$_2$O$_4$	TeO$_2$	SnO$_2$	MoO$_2$	Fe$_3$O$_4$	CoO/Co$_3$O$_4$
Formation of aldehydes	0·19	0·45	0·42	0·46	0·17	0·29
Formation of CO$_2$	0·19	0·56	1·14	2·78	4·36	21·2
Consumption of C$_4$H$_8$	0·29	0·74	0·84	1·63	1·5	6·0

Table 30. Activation energies and reaction orders with respect to isobutene and oxygen, for the formation of certain oxidation products [630]

Catalyst	Order w.r.t. O$_2$			Order w.r.t. iC$_4$H$_8$			E (kcal. mole^{-1})		
	alds.	CO$_2$	CO	alds.	CO$_2$	CO	alds.	CO$_2$	CO
Sb$_2$O$_4$	1	1	1	0·6	0·1	0·52	12	22	13
TeO$_2$	0	0	0	0·6	0·6	0·6	26	35	27
CoO/Co$_3$O$_4$	—	1	—	—	0	—	—	20	—
Fe$_3$O$_4$	—	0·7	—	—	0	—	—	18	—
MoO$_2$	0·5	0·5	0·4	0·6	0·6	0·5	21	15	12
SnO$_2$	0·8	0·8	0·87	0·65	0·25	0·40	17	25	21
Sb : Te(4 : 1) + 60 at.% Co†	0·6	0·6	0·7	0·75	0·85	0·50	23	28	30
Sb : Te(4 : 1) + 60 at.% Fe	0·5	0·3	0·4	1·0	0·9	0·8	20	19	26
Sb : Te(4 : 1) + 60 at.% Mo‡	0·57	0·47	0·70	0·60	0·47	0·55	20	21	31
Sb : Te(4 : 1) + 80 at.% Sn§	0·50	0·45	0·50	0·80	0·80	0·75	19	23	30
Sb : Te(4 : 1)	0·4	0·4	0·5	0·68	0·70	0·57	20	35	30
Cu$_2$O‖	—	—	—	—	—	—	15	20	—

† Studied further in Ref. [632].
‡ Most selective catalyst. Studied further in Ref. (634).
§ Activity due to the formation of a solid solution of Sb$_2$O$_4$ in SnO$_2$
‖ Ref. [632].

In other studies, Malinowski [635] for example, investigated the oxidation of isobutene over various bismuth molybdate catalysts, and found that the optimum conversion (88·9 %) of isobutene and selectivity (92·1 %) to methacrolein, was achieved by passing an isobutene–oxygen–steam (1 : 1·9 : 4·5) mixture over a catalyst consisting of Bi$_2$O$_3$ · 2MoO$_3$. Zhiznevskii *et al.* [636] had earlier investigated the effect of Bi$_2$O$_3$–MoO$_3$–P$_2$O$_5$ (1·2 : 1 : 0·0042) catalysts, supported (20 % loading) on silica gel, alumina, pumice and porous glass, in the oxidation of isobutene at temperatures between 400 and 545°C. The maximum yield of methacrolein (28 % isobutene converted to products containing 72 % methacrolein) was obtained, at 530°C, over a porous glass-supported catalyst, using a 1 : 4·8 : 1 isobutene : air : water mixture.

Kakinoki *et al.* [637, 638] have studied V_2O_5–Li_2SO_4 catalysts supported on various carriers, including anatase, gypsum, celite and silica gel. Oxygen gas was adsorbed on the catalysts at 470°C, its uptake was measured, and Langmuir adsorption isotherms were obtained. The selectivity towards methacrolein in this system depended on the amount of oxygen adsorbed/unit surface area of catalyst. Kakinoki *et al.* [639] also examined V–Mo–P catalysts supported on identical carriers $(V_2O_5:MoO_3:P_2O_5:carrier = 12:6:1\cdot2:80\cdot8)$ in the temperature range 305–460°C, using both air and air/nitrogen mixtures as oxidants. Comparisons of the selectivity obtained for the various oxidants led to the conclusion that the formation of carbon dioxide occurred simultaneously with methacrolein formation and was unconnected with the concentration of oxygen in the gas-phase, although carbon monoxide originated mainly from the further oxidation of methacrolein.

Finally, Moro-oka *et al.* [640] investigated the catalytic properties of various oxides in the oxidation of isobutene, acetylene and propane and correlated these with the ratio of the heat of formation of the oxide to the number of oxygen atoms in the oxide molecule (ΔHo). A relationship was found between the catalytic activity and ΔHo: the higher the activity, the lower ΔHo. For isobutene and acetylene, as ΔHo increased, the reaction order with respect to hydrocarbon increased, but the order in oxygen decreased. It was also believed that the hydrocarbon reacts via the adsorbed state since the sequence of adsorption strength, viz.

isobutene > acetylene > propylene > ethylene > propane

was the reverse of the reaction-order sequence.

Apart from being studied by Serban *et al.* [641], the ammonolysis of isobutene to methacrylonitrile has been rather neglected. According to Serban *et al.* [641] two different reactions are possible during this process, viz. (i) the endothermic reaction of isobutene with ammonia and (ii) the exothermic reaction between the two, which occurs in the presence of oxygen. A theoretical appraisal was made of the thermodynamics and kinetics of process (i), and a series of curves were obtained relating the equilibrium conversion of isobutene and the temperature, for values of the isobutene : ammonia mole ratio between one and ten. From these data, it was predicted that, with an isobutene : ammonia ratio of 1 and temperatures in excess of 620°C, for example, a 70% conversion of isobutene to methacrylonitrile would be obtained. It was further predicted that an increase in the ammonia : isobutene ratio would result in the same conversion taking place at lower temperatures. Serban *et al.* [641] investigated the oxidative ammonolysis (reaction (ii)), using silica gel-supported bismuth molybdate catalysts. Although the oxidation of isobutene proceeded at temperatures between 350 and 360°C, temperatures in excess of 380°C were required for the appearance of methacrylonitrile. Over the temperature range 390–430°C, despite a sharp increase in the consumption of isobutene, the selectivity of the process towards methacrylonitrile decreased; from 460 to 480°C, little variation was observed either in selectivity or catalytic activity. From the kinetic data thus obtained, Serban et al. [641] derived activation energies for methacrylonitrile formation (20·4 kcal. mole^{-1}; 390–460°C) and the oxidation of isobutene to carbon dioxide (26·5 kcal. mole^{-1}). An activation energy of 21·2 kcal. mole^{-1} was also obtained for the formation of hydrogen cyanide from isobutene.

4.7. To Other Compounds

Although these will not be dealt with in great detail, isobutene undergoes other reactions very similar to those observed with propylene. For example, just as propylene and acrolein may be catalytically oxidized to acrylic acid, so isobutene and methacrolein yield methacrylic acid under similar conditions [331, 642, 643]. Furthermore, under conditions where propylene undergoes dehydroaromatization to benzene and hexa-1,5-diene [438], isobutene yields p-xylene and 2,5-dimethyl-hexa-1,5-diene. For example, Ohdan et al. [438] obtained a 61·5% yield of p-xylene from isobutene, using Sn–Bi (1:1) catalysts at 470°C. Extremely high yields (up to 71·9%) of 2,5-dimethyl-hexa-1,5-diene have also been obtained in the presence of SnO$_2$–MoO$_3$ (9:1) catalysts at temperatures between 136 and 190°C [228]. Finally, using the same catalysts, yields of 2-methyl-propan-2-ol of between 83 and 100%, have been obtained in the low temperature (90–105°C) oxidation of isobutene.

CHAPTER 5

The Catalytic Oxidation of C_5-Hydrocarbons

5.1. General Introduction

Despite the title of this chapter the major part of it will be devoted to C_5-olefins and diolefins, since industrial interest in pentenes and C_5-hydrocarbons in general, lies in the fact that, under certain conditions, they can be oxidatively dehydrogenated to yield diolefins and in particular, isoprene.

The pentenes may be divided into two groups:

    (1) straight-chain pentenes; and

    (2) branched-chain pentenes.

There are three straight-chain pentenes (pent-1-ene and *cis*- and *trans*-pent-2-ene) and three branched-chain pentenes (2-methyl-but-2-ene, 2-methyl-but-1-ene and 3-methyl-but-1-ene). The latter compounds are often referred to by the trivial names, isopentenes or iso-amylenes. Straight-chain pentenes may be catalytically dehydrogenated to yield penta-2,4-diene (piperylene) and iso-pentenes may be dehydrogenated, yielding 2-methyl-buta-1,3-diene (isoprene).

The industrial interest in isoprene lies in the fact that it can be converted, by stereospecific polymerization, into polymeric material identical with natural rubber, or, co-polymerized with, say, isobutene to yield a product which can be vulcanized [644]. Synthetic elastomers such as *cis*-1,4-polyisoprene rubber often have properties superior to those of natural rubber [645, 646] with respect to homogeneity and resistance to cold and ageing, and are much used in the manufacture of tyres. Isoprene may also be oxidized in the vapour phase (300–450°C), over V–Mo or U–As catalysts, to give citraconic acid [647].

Nowadays, isoprene can be made industrially by the oxidation of a mixed C_5-cut or refinery stream without any pre-separation. The composition of such a fraction is shown below [54]:

C_5-component	W%
n-pentane	11·27
3-methyl-but-1-ene	18·32
2-methyl-but-1-ene	12·27
2-methyl-but-2-ene	27·79
pent-2-ene	21·21

For example, a mixed C$_5$-stream obtained from the thermal distillation of crude oil, fluidized catalytic cracking, naphtha cracking or olefin disproportionation, may be treated at 371–704°C in the presence of, say, a Fe–P–Re (29%, 20%, 6·3%) oxide catalyst, to dehydrogenate the isopentanes, without affecting the straight-chain pentenes and pentanes or branched-chain pentanes [648]. A second method of preparation involves the vapour-phase reaction of isobutene with formaldehyde in the presence of catalysts such as boric acid-silica gel [649], Sb$_2$O$_3$–silica gel [650], or silica gel impregnated with lead nitrate and phosphoric acid and heated to 300–350°C [651]. Thus, in the presence of a boric acid–silica gel catalyst, it is claimed that formaldehyde–isobutene mixtures (1:3 to 1:8) are converted to isoprene with yields of between 68 and 43%.

Before dealing in detail with the oxidation of the pentenes, it may be instructive to compare the reactivities of these compounds with hydrocarbons previously discussed (propylene, butene). By feeding mixtures of alkenes and but-1-ene over bismuth catalysts at 460°C, Voge and Adam [3, 655] obtained the reactivities of some pentenes. These are shown in Table 31. Similarly, Belen'kii et al. [652] oxidized branched-chain pentenes in the presence of bismuth molybdate catalysis (3:7 Bi–Mo 4:6) and found that the reactivities lay between those of but-1-ene (the most reactive) and propylene (the least).

Table 31. Relative reactivities of some olefins in the presence of Bi–Mo–O catalysis at 460°C

Olefin	Rel. Reactivity/molecule	No. of type of allylic H-dtons	Reactivity /H atom
But-1-ene	1·00	2 (s, s, p)†	0·50
Pent-1-ene	1·38	2 (s, s, p)	0·69
2-methyl-but-1-ene	4·2	2 (s, t, p) + 3 (p, t, p)	—
Propylene	0·11	3 (p, s, p)	0·037

† Letters in brackets—first letter denotes type of C atom holding the allyl H; second and third letters denote type of nearest and furthest vinyl C.
p = primary; s = secondary; t = tertiary.

5.2. Straight-Chain Hydrocarbons

A. Copper Oxide-Containing Catalysts

Unlike the oxidation of propylene, the oxidation of pent-1-ene and pent-2-ene in the presence of cuprous oxide-containing catalysts is unselective, yielding a large number of products [653]. For instance, Gorokhovatskii and co-workers [653] oxidized a mixture of pent-1-ene and pent-2-enes (59:41 v/v) in the presence of carborundum-supported copper(I) oxide (14·9 mg Cu/cm^3; carborundum grain-size, 0·25–0·50 mm) at 280–360°C. The products obtained included pent-2-en-4-one, pent-1-en-3-one, pent-2-enal (trace), propionaldehyde, acrolein, penta-2,4-dien-5-al, 1,3- and 2,3-pentadienes and small amounts of C_1 to C_5 saturated aldehydes and ketones, including methyl ethyl ketone and acetone. Vovyanko and Gorokhovatskii [654] also studied the oxidation of pent-2-ene and pentane over copper oxide at 320°C, but reported carbon dioxide and water as the major products.

B. Bismuth Oxide–Molybdenum Oxide

In 1965, Adams [655] reported the results of studies on the oxidation of a large number of olefins including pent-1-ene and 2-methyl-but-2-ene, over bismuth molybdate catalysts at 460°C. At 5% conversion, pent-1-ene was oxidized to penta-1,3-diene with a selectivity of 88%, but this fell to 79% at 15% hydrocarbon conversion. In contrast to this behaviour, but-1-ene, under identical conditions, yielded buta-1,3-diene with a 95% selectivity at 20% conversion and almost the same selectivity (90%) at 80% conversion. It is apparent that, at high conversions, the oxidation of olefins containing more than four carbon atoms suffers a loss of selectivity. The main cause of the phenomenon at high hydrocarbon conversion seems to be inhibition by the initial reaction by-products.

In previous chapters it was seen that oxidation of propylene and butenes, in the presence of a bismuth molybdate catalyst, obeys first-order kinetics with respect to the alkene and that the rate is independent of oxidation products. With higher olefins, such as pent-1-ene, this, apparently, is no longer the case. Adams [655] studied the oxidation of pent-1-ene using a flow reactor and investigated the influence of olefin, oxygen and pendadiene concentrations. The rate of oxidation was found to fit the following expression:

$$\text{Rate} = \frac{k[\text{alkene}]K_1^{0.5}[O_2]^{0.5}}{1 + K_1^{0.5}[O_2]^{0.5} + K_2[\text{diene}]}$$

At 460°C, $K_1^{0.5} = 21$ atm.$^{-0.5}$ and $K_2 = 700$ atm.$^{-1}$

According to Adams [655], the mechanistic significance of the equation is that the rate-controlling step involves the abstraction of an allylic hydrogen atom from a gas-phase or physisorbed olefin molecule by oxygen chemisorbed on the catalyst. Reaction products, such as pentadiene, interfere with oxygen chemisorption.

C. Other Catalysts

Tan et al. [228] found that, in the presence of Co_3O_4–MoO_3 or SnO_2–MoO_3 (Co or Sn:Mo = 9:1), pent-1-ene yields a mixture of methyl propyl ketone and diethyl ketone. Thus, when pent-1-ene (3 v%)–oxygen (30 v%) mixture was passed, with steam (30 v%), over a 9:1 SnO_2–MoO_3 catalyst, the following products were observed:

Temp. (°C)	158	185
Conversion (%)	8·0	22·0
S^y to C_5-ketones† (%)	65·4	52·6
$4/5C_3H_7CO_2H$	1·0	1·4

† C_5-ketones = methyl propyl ketone and diethyl ketone.

The oxidation of a number of olefins over supported iridium catalysts, has been investigated by Cant and Hall [656]. The two pent-2-enes were oxidized at 183°C in the presence of 5% Ir-silica and 5% Ir-alumina catalysts. The oxidation products included acetic acid (selectivity 32%), propionic acid (selectivity 8%) and acetone (selectivity < 0·1%). Some saturated and unsaturated C_5-ketones were also detected.

Finally, Butt and Fish [657] have investigated the oxidation of pent-1-ene and cis- and trans-pent-2-ene over freshly-prepared and "old" pumice-supported V_2O_5 catalysts. Cis- and trans-pent-2-ene behaved identically in the reaction and the oxidation of pent-1-ene proceeded largely by isomerization to pent-2-ene, followed by reaction of the latter. "New" and "old" (i.e. used for a period exceeding 5 hours) catalysts showed vastly different behaviour. The oxidation of pent-2-ene at 250–450°C, over a freshly-prepared catalyst, yielded acetaldehyde as the major oxygenated product, together with propionaldehyde, ethylene, propylene, trans-2,3-epoxypentane, acetone and carbon dioxide. After 5 hours' use, the character of the oxidation changed becoming more rapid and producing fourteen more products, including methanol, n-butyraldehyde, butanone, pentan-2-one, pentan-3-one, small amounts of C_{10}-hydrocarbons, crotonaldehyde, ethanol, butan-2-ol, ethylene oxide, propylene oxide, 2-methyl-2,3-epoxypropane, t-2,3-epoxybutane, and 3-methyl-butan-2-one. It was suggested that on a new catalyst the products are probably formed by the interaction of a surface-bonded C_5-cation with O^{2-}:

$$CH_3CH{=}CH{-}CH_2{-}CH_3 \xrightarrow[e^-\text{ transfer}]{\text{Adsorption and}} CH_3\overset{+}{C}H{-}CHCH_2CH_3 \quad \text{I}$$

$$\text{and } CH_3CH{-}\overset{+}{C}HCH_2CH_3 \quad \text{II}$$

Then:

$$I + O^{2-}(ads.) \longrightarrow CH_3\overset{\overset{O^{\underline{2}}}{|}}{C}{-}CH{-}CH_2{-}CH_3 \quad \text{III} \longrightarrow CH_3CHO + {}^-\overset{\curvearrowleft}{C}HCH_2CH_3$$

$$\begin{array}{c} CH_3 \quad\quad C_2H_5 \\ \diagdown \quad\quad\quad \diagup \\ HC{-----}CH \\ \diagdown \quad\quad \diagup \\ O \end{array} \xleftarrow[\text{closure}]{\text{ring}} \begin{array}{c} \overset{\overset{O^{\cdot}}{|}}{} \\ CH_3C{-}CH{-}CH_2{-}CH_3 \end{array} \quad \xleftarrow{-e^-}$$

$$CH_2{=}CH{-}CH_3$$

$$\text{II} + \text{O}^{2-}(\text{ads.}) \longrightarrow \underset{\underset{\wedge\wedge\wedge}{\overset{\overset{\text{O}^-}{|}}{}}}{\text{CH}_3\text{CH}-\text{CH}-\text{CH}_2-\text{CH}_3} \quad \text{IV} \xrightarrow{\text{similarly}}$$

propionaldehyde, ethylene, 2,3-epoxypentane

In the presence of "old" catalysts, a different mechanism was considered. Without gaseous oxygen, no oxidation occurred when pent-2-ene was passed over such catalysts, suggesting that this was indeed the active oxidant. Under these conditions, the oxidation acquired a free radical-chain character, which explained the sudden loss of selectivity:

$$\text{CH}_3-\text{CH}=\text{CH}-\text{CH}_2-\text{CH}_3 \xrightarrow{\text{O}_2} \underset{\overset{|}{\text{OO}^{\cdot}}}{\text{CH}_3-\text{CH}-\text{CH}-\text{CH}_2-\text{CH}_3}$$

$$\text{CH}_3-\text{CHO} + \text{C}_2\text{H}_5\text{CHO}$$

$$\uparrow$$

$$\underset{\overset{|\quad|}{\text{O}-\!-\text{O}}}{\text{CH}_3-\text{CH}\!-\!\text{CH}-\text{CH}_2-\text{CH}_3}$$

$$\underset{\overset{\text{O}-\text{O}}{}}{\text{CH}_3-\text{CH}\text{,}\text{CH}-\text{CH}_3} \quad \overset{\text{CH}_2\text{,}}{}$$

$$\underset{\overset{|}{\text{OO}^{\cdot}}}{\text{CH}_3-\text{CH}-\text{CH}-\text{CH}_2-\text{CH}_3} \; \underset{\xrightarrow{\text{isomerization}}}{\xrightarrow{\text{ring-closure}}}$$

$$\downarrow$$

$$\text{CH}_3-\text{CHO} + \text{CH}_3\overset{\overset{\text{O}}{\|}}{\text{C}}-\text{CH}_3$$

The formation of large amounts of methanol under unselective conditions was most easily accounted for on the basis of a "hydroperoxylation mechanism" [658]:

$$\text{CH}_3-\text{CH}=\text{CH}-\text{CH}_2-\text{CH}_3 \xrightarrow[\text{by catalyst}]{\text{H-abstraction}}$$

$$\text{CH}_3-\text{CH}=\text{CH}-\overset{\cdot}{\text{CH}}-\text{CH}_3 \xrightarrow{\text{O}_2} \underset{\overset{|}{\text{OO}^{\cdot}}}{\text{CH}_3-\text{CH}=\text{CH}-\text{CH}-\text{CH}_3}$$

$$\underset{\overset{|}{\text{OO}^{\cdot}}}{\text{CH}_3-\text{CH}=\text{CH}-\text{CH}-\text{CH}_3} \xrightarrow{\text{RH}} \underset{\text{H}}{\overset{\text{CH}_3}{}}\text{C}=\text{C}\underset{\text{O}\!-\!\text{OH}}{\overset{\text{CH}_3}{}} \longrightarrow$$

$$\text{CH}_3\text{OH} + \text{CH}_3-\text{CH}=\text{CH}-\text{CHO}$$

Butt *et al.* [659] also investigated the changes in surface area and phase-composition of vanadia catalysts in contact with reacting mixtures of pent-2-ene and oxygen. With pumice-supported vanadia, vanadium(V) was gradually reduced to vanadium(IV). When vanadium(IV) reached a critical concentration, a phase-change occurred and a reaction, involving the support, giving sodium vanadyl vanadate, took place, This reaction coincided with a sudden increase in the surface area of the catalyst and the overall rate of pentene oxidation, but a decline in the selectivity of the reaction.

5.3. Branched-Chain Hydrocarbons

A. Isopentane

The direct oxidation of isopentane to isoprene has been the subject of a number of studies by Russian workers. Kolobikhin [660], for example, carried out the conversion using iodine in the presence of hydrogen iodide-acceptors (MnO$_2$, CaO/CaCO$_3$ and NaOH on alumina). Under optimum conditions (520°C; $t_c = 2\cdot4$ s; O$_2$:i-pentane:I$_2$ = 1:1:0·017), 50% isoprene was obtained with a selectivity of 65%. Sakovich [661], Adel'son and co-workers [662] and Skarchenko *et al.* [663] have investigated surface effects in the dehydrogenation of isopentane with iodine at temperatures in the region of 500°C. Adel'son *et al.* [662], for example, examined the influence of quartz, glass and KCl or KI on quartz and determined the isoprene yields.

Stergilov *et al.* [664,665] have also used bromine and hydrogen bromide in the reaction. Stergilov *et al.* [664] tested the effect on the reaction of a number of catalysts, including silica, cobalt(II) oxide/silica, cobalt(II) oxide on pumice, cobalt(II) chloride/pumice, lithium chloride/pumice, magnesium chloride/pumice, cerium(III) chloride/pumice and rear-earth chlorides/pumice. A catalyst consisting of 1:1:1 CoCl$_2$–LiCl–MgCl, supported on pumice, gave the best results. With this catalyst, a iC$_5$H$_{12}$:O$_2$:HBr:H$_2$O (1:1:2:29) mixture gave optimum conversion (41–49%) and a 24–28% yield of isoprene at 500°C. Similar yields of isoprene were also obtained at 500°C using a feed containing 0·1 mole bromine and 1·6–2 mole oxygen.

B. Isopentenes

Due to their greater reactivity, the dehydrogenation of isopentenes to isoprene is a more satisfactory process than the above, and several patent claims indicate that isoprene can be prepared, with good selectivity, at fairly high levels of isopentene conversion. A range of catalysts may, apparently, be used, including Na or Ca type A molecular sieves [666], unmodified or boric acid-modified ferrites of manganese, nickel or zinc [667], or alumina containing sodium ions [668]. A catalyst containing antimony and molybdenum oxides [669] is particularly good, 50% of an isopentene mixture being converted to isoprene at 400°C with the following selectivities: 3-methyl-but-1-ene (94·3 %), 2-methyl-but-1-ene (93·4 %) and 2-methyl-but-2-ene (91·1 %).

The following sections discuss studies of the oxidation of isopentenes which have been reported outside the patent literature.

(a) *Copper oxide-containing catalysts*. Studies of the oxidation of isopentene isomers in the presence of copper-containing catalysts have been undertaken mainly by the Russian workers [545, 654, 670–673]. A number of reactions appear to take place including oxidative dehydrogenation to yield isoprene, oxidation yielding C_5- and lower carbon-number carbonyls and complete oxidation [672]. Depending on the type of reaction occurring, isopentenes vary in reactivity. Thus, at temperatures in the range 275–350°C. Belen'kii *et al*. [673] investigated the reactivity, not only of the isopentenes, but also of but-1-ene and but-2-ene. As regards oxidative dehydrogenation, the following order of reactivity was obtained:

3-methyl-but-1-ene > but-1-ene > 2-methyl but-1-ene > 2-methyl-but-2-ene > but-2-ene.

(cf. Vovyanko and Gorokhovatskii [654] where the order was found to be: 3-methyl-but-1-ene > 2-methyl-but-2-ene > 2-methyl-but-1-ene.) The yield of isoprene from 2-methyl-but-1-ene was about double that obtained from 2-methyl-but-2-ene [545, 673], whilst the highest yield of all was obtained from 3-methyl-but-1-ene. Reactivity towards partial oxidation (presumably to carbonyl compounds), declined in the order:

2-methyl-but-1-ene > 2-methyl-but-2-ene ≫ 3-methyl-but-1-ene.

Detailed studies of the products obtained during the oxidation of the isopentenes have been made by Gorokhovatskii *et al*. [654, 672]. Thus Gorokhovatskii *et al*. [672] investigated the oxidation of various mixtures of isopentenes in a flow system at temperatures between 314 and 361°C. The main oxidation products were carbon dioxide, isoprene, acetone, ethylacrolein and methyl isopropenyl ketone. Acrolein, methylketene, 3-methyl-but-4-al-1-ene and 2-methyl-but-4-al-1-ene were formed in small amounts and traces of isovaleraldehyde, 3-methyl-but-4-al-2-ene were also produced. With increasing temperature and contact time, the yields of carbon dioxide, ethylacrolein, methyl isopropenyl ketone and the C_3-carbonyl compounds increased. It was thought that ethylacrolein and methyl isopropenyl ketone were probably formed from 2-methyl-but-1-ene and 2-methyl-but-2-ene respectively.

Vovyanko and Gorokhovatskii [654] examined the products of oxidation of the individual isoprenes. At 320°C, 2-methyl-but-1-ene yielded isoprene, C_5-aldehydes, 2-methyl-but-1-ene, carbon dioxide and water, whilst the last two compounds were the sole products of the oxidation of 3-methyl-but-1-ene.

Finally, although not strictly relevant to this section, Popova *et al*. [670] have reported that copper(I) oxide catalysts containing the oxides of certain heavy metals could be used for the selective oxidation of pentadienes. Thus, in

the oxidation of 2-methyl-buta-1,3-diene and penta-1,3-diene at 390°C, 75%
of the corresponding dien-al was obtained.

(b) *Bismuth oxide–molybdenum oxide catalysts.* In the presence of bismuth
molybdate catalysts, isopentenes undergo isomerization, oxidative dehydrogenation to yield isoprene and also some oxidation to yield C_5-unsaturated
aldehydes. Aliev *et al.* [674] found that, at 400°C, the oxidation of an isopentene
mixture (14·5% 3-methyl-but-1-ene, 60·0% 2-methyl-but-2-ene, 25·5% 2-
methyl-but-1-ene) to yield isoprene, was first-order with respect to the pentenes
and the rate was independent of the oxygen concentration. The following rate
constants and activation energies for dehydrogenation were also obtained
for the individual pentenes:

Pentene	$k(s^{-1})$	E_a (kcal. mole^{-1})
3-methyl-but-1-ene	0·95	18·9
2-methyl-but-2-ene	0·80	18·6
2-methyl-but-1-ene	0·55	20·0

Activation energies for the extensive oxidation and isomerization of the
mixture were 22·0 and 4·0–5·0 kcal. mole^{-1} respectively.

Adams [655] examined the effect of structure on the reactivities of the isopentenes, at 460°C, in the presence of bismuth molybdate. The following results
were obtained (taken from Ref. [175]).

Olefin	Rel. Reactivity/molecule	No. and type of allylic H atoms	Reactivity /H atom
2-methyl-but-2-ene	2·7	6 (p, t, s) + 3 (p, s, t)	0·30
2-methyl-but-1-ene	4·2	2 (s, t, p) + 3 (p, t, p)	—
3-methyl-but-1-ene	2·7	1 (t, s, p)	2·7

p = primary; s = secondary; t = tertiary.

In the same paper, Adams [655] also reported the selectivities of the isopentenes towards isopentene and C_5-unsaturated aldehydes. These are shown
in Table 32 and but-1-ene is included for comparison.

This table shows that the selectivity towards isoprene, although initially
fairly high, declines rapidly at higher conversions, particularly in the case of
2-methyl-but-1-ene and 2-methyl-but-2-ene. This variation in selectivity is not
a function of the reactivity of the olefins but, as with the straight-chain pentenes,
is caused by inhibition by reaction products. The severity of inhibition also
depends to a large extent on the structure of the product. With isopentenes,

Table 32. Selectivities towards various products during the oxidation of the isopentenes over bismuth molybdate at 460°C

Olefin	Conversion (%)	Product selectivity (%)		
		Diene	C_5H_6O	C_5H_8O
3-methyl-but-1-ene	40	80†	4	—
2-methyl-but-1-ene	35	60†	3	10
	40	58†	n.a.	n.a.
2-methyl-but-2-ene	20	73†	4	9
	40	62†	9	9
	70	44†	16	9
but-1-ene	20	95‡	—	—
	40	95‡	—	—
	80	90‡	—	—

† Isoprene.
‡ buta-1,3-diene.   n.a. = not available.

possible reaction products are conjugated dienes and unsaturated aldehydes; isoprene is only a moderately good inhibitor, while unsaturated aldehydes are apparently very strong inhibitors [675]. Adams [655] showed that, when structurally allowable, conjugated dienes are the preferred products in olefin oxidation rather than unsaturated aldehydes. However, the initial products may also react further if additional conjugated systems can be formed. In general products from a given olefin can be predicted quite well by considering the initial attack on the molecule to involve removal of a hydrogen atom from an allylic position. Dehydrogenation can then occur by the abstraction of an adjacent hydrogen atom, thus forming a conjugated diene. If a methyl group in the molecule has the requisite structure (i.e. attached to a vinylic carbon), aldehydes may also be formed.

Adams [675] obtained a crude estimate of the amount of inhibition caused by reaction products, by measuring the rate of reaction occurring in an integral reactor and comparing it with the rate calculated from molecular reactivities. These values are shown in Table 33. Unsaturated aldehydes are much more strongly adsorbed (ca. 2 orders of magnitude) than the corresponding diene [675]. Thus, the oxidation of 3-methyl-but-1-ene, which produces only isoprene as an initial product, should be only moderately strongly inhibited until appreciable conversion allowed the formation of the diene aldehyde, and the consequent onset of severe inhibition. The oxidation of 2-methyl-but-1-ene, should however be severely inhibited from the start. This is well-illustrated in Fig. 2 where the upper two lines represent the course of the reaction without inhibition.

A reaction scheme for the oxidation of the isopentene isomers is shown in Fig. 3.

Table 33. Relative rate of inhibited and uninhibited reaction for olefin oxidation over bismuth molybdate at 460°C

Integral rate when product partial pressure at the reaction exit = 0·015 atm. and p_{O_2} = 0·12 atm. (taken from Ref. [675]).

Olefin	Ratio inhib. rate/uninhib. rate
pent-1-ene	0·4
pent-2-ene	0·4
3-methyl-but-1-ene	0·5
2-methyl-but-1-ene	0·04

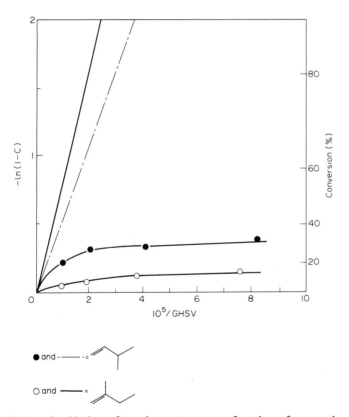

FIG. 2. Rates of oxidation of two isopentenes as a function of conversion (from Ref. [675]).

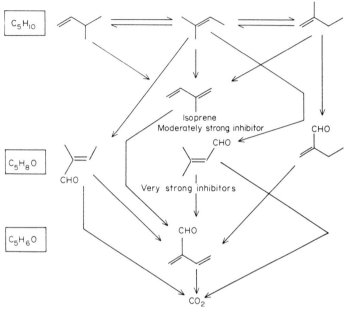

FIG. 3. Reaction scheme for the dehydrogenation of isopentene isomers.

Finally, other studies involving bismuth molybdate catalysts have been carried out by Watanabe *et al.* [676] and Gusman *et al.* [677]. Watanabe *et al.* [676] oxidized a mixture of isopentenes (3-methyl-but-1-ene: 2-methyl-but-1-ene: 2-methyl-but-2-ene = 2:25:73) in a pulse-flow microreactor and obtained the highest yield of isoprene from 3-methyl-but-1-ene (activation energy, 340–420°C = 42 kcal. mole^{-1}). The 2-methyl-butenes were found to isomerize more quickly than 3-methyl-but-1-ene.

Gusman *et al.* [677] attempted to develop an effective catalyst for the oxidative dehydrogenation reaction based on 10–13% bismuth molybdate supported on silica gel, borosilicate- and aluminosilicate-material. Between 430 and 480°C, and at constant pore radius, the specific surface area of the support had no effect on the kinetics. However, an increase in the pore radius markedly increased the degree of isopentene conversion. Increasing the pore radius beyond 50 Å did not alter the selectivity of the catalyst.

(*c*) *Other catalysts.* Gagarin *et al.* [678] have examined the catalytic dehydrogenation of isopentenes over a variety of oxide catalysts (Bi–Mo, Sn–Sb, Fe–Mo, Fe–Zn–Cr). Rate constants were obtained which are shown below (Table 34).

Table 34.

Olefin	$k(s^{-1})$			
	Bi–Mo	Sn–Sb	Fe–Mo	Fe–Zn–Cr
3-methyl-but-1-ene	0·80 (0·95)†	0·50	0·40	0·15
2-methyl-but-1-ene	0·50 (0·55)†	0·45	0·50	0·30
2-methyl-but-2-ene	0·45 (0·80)†	0·20	0·60	0·45

( )† = Results of Aliev, Gagarin, Yanovskii and Zhomov [674].

Nishikawa *et al.* [679] also examined the oxidative dehydrogenation of an isopentene mixture (80% 3-methyl-but-1-ene, 10% 2-methyl-but-2-ene, 10% 2-methyl-but-1-ene) in the presence of various binary oxide catalysts (U–Sb, Sn–Sb and Bi–Mo) at various temperatures. At 400°C and low conversions, the selectivity towards isoprene was high with all three catalysts, but, as the conversion increased, a marked drop in selectivity was noted, except for the U–Sb–O catalyst. However, with U–Sb–O, the selectivity also declined if the oxidation temperature exceeded 400°C. With both U–Sb–O and, surprisingly, Bi–Mo–O, the partial pressure of oxygen in the reaction mixture affected selectivity.

Phosphoric acid-containing catalysts have been recently investigated by Usov *et al.* [680–682]. Usov [682] initially observed the reaction over catalysts containing 5, 18 and 30 wt% phosphoric acid, supported on silica, at temperatures between 350 and 450°C. The reaction was first-order with respect to the isopentenes and zero-order with respect to oxygen throughout the temperature range investigated. At 400°C, the apparent rate constants were 0·52, 0·89 and 0·72 cm³/s-g cat. with increasing phosphoric acid content, and activation energies of 20·0, 16·3 and 18·3 kcal. mole⁻¹, respectively, were obtained.

Further investigations by Usov *et al.* [681] revealed that isomerization of the isopentenes occurred to a much greater extent than oxidation and in the reaction

$$\text{2-methyl-but-1-ene} \underset{k_r}{\overset{k_f}{\rightleftharpoons}} \text{2-methyl-but-2-ene}$$

$k_f = 2$–$2·4 \times k_r$ in the temperature range 375–425°C. In their latest paper, Usov *et al.* [680] determined the total concentration of adsorptive and catalytic centres on the surface of phosphoric acid-containing catalysts by impulse poisoning with butylamine and concluded that oxidative dehydrogenation of isopentenes proceeded via an intermediate complex of an olefin carbonium ion and chemisorbed oxygen.

Butt and Fish [683] have examined the vanadium pentoxide-catalyzed oxidation of each of the three branched-chain pentenes in the temperature range 200–400°C. In their experiments, Butt and Fish often employed an extremely long contact time (33·6 s) and an oxygen:olefin ratio of 1·25. Under these conditions, the V_2O_5-catalyzed oxidation of 2-methyl-but-2-ene was studied over a range of temperatures. At 220°C, more than 20 % of the olefin was converted to 2-methyl-but-1-ene but no oxygenated products were observed; at 240°C, the oxidation products included acetaldehyde, acetone, 2,3-epoxy-2-methylbutane, 3-methyl-butan-2-one and unspecified C_{10}-hydrocarbons. The highest yield of 2,3-epoxy-2-methylbutane (13 %, based on pentene introduced) was obtained at 300°C, whilst the highest yields of acetone (43·5 %) and acetaldehyde (62·5 %) were obtained at 350°C. Qualitatively, the catalyzed oxidation of 2-methyl-but-2-ene gave products very similar to those of its gas-phase oxidation; in fact, the latter process appeared to be more selective. Thus, an 85·5 % yield of acetaldehyde and a 50 % yield of acetone were obtained at 300°C. The yield of acetone even increased to 90 % as the temperature was raised to 350°C.

2-methyl-but-1-ene yielded products qualitatively similar to those from 2-methyl-but-2-ene, in the presence of vanadium pentoxide, although considerable isomerization to the latter compound also occurred. Although more resistant to oxidation, 3-methyl-but-1-ene similarly produced large amounts of 2-methyl-but-2-ene and oxygenated products characteristic of the latter olefin. A comparison of the products of the three isopentenes is shown in Table 35.

Table 35. Products of the V_2O_5-catalysed oxidation of the isopentenes† [683]

Product	Yield (mole % based on fuel introduced)		
	2-methyl-but-2-ene	2-methyl-but-1-ene	3-methyl-but-1-ene
2-methyl-but-2-ene	14·5r	15·3	6·9
2-methyl-but-1-ene	6·5	7·0r	0·5
3-methyl-but-1-ene	0	0	63·5r
C_{10} hydrocarbons	3·5	10·0	1·8
ethylene	3·0	2·5	tr.
propylene	2·0	2·0	tr.
acetaldehyde	40·0	37·0	2·7
acetone	32·5	31·0	4·0
2-methyl-butan-2-one	7·5	tr.	0·3
butanone	1·0	8·0	1·0
2,3-epoxy-2-methylbutane	13·0	8·0	0·5

† O_2 : pentene = 1·25; t_c = 33·6s; Temp. = 300°C. r = amount of pentene unconverted. tr. = traces.

It was thought probable that, during oxidation, each pentene isomerized to give 2-methyl-but-2-ene and the products observed were characteristic of the latter compound. The species O^{2-} was postulated to be the oxidant rather than gaseous oxygen, yielding surface-bonded anions after reaction with adsorbed olefin:

Products such as acetone, acetaldehyde and olefins were thought to be characteristic of the following scission reactions:

The presence of unequal amounts of acetone and acetaldehyde in the reaction products demonstrates that adsorption gives non-equivalent amounts of cations I and II.

Finally, if the adsorbed anions III and IV lose an electron to the surface the C—C bond breaks, the resulting adsorbed alkoxy radicals may rearrange giving 2,3-epoxy-2-methylbutane and 3-methyl-butan-2-one:

CHAPTER 6

Mechanism of Selective Oxidation of Hydrocarbons

6.1. Generalized Theories of Catalysis

Fundamental research into catalysis is justified, commercially, by the hope that knowledge thus acquired will assist in the development of catalysts which are cheaper, more active and more selective than those hitherto in use [684]. The ability to predict catalytic activity and selectivity would, of course, be enormously helpful in the selection of potential catalysts and this problem has been, and will remain, a very challenging one.

Notable theories have been put forward correlating many reactions and many catalysts and these include the electronic theory of catalysis [685], the "multiplet theory" of Balandin [686] and the "crystal field theory" of Dowden [687, 688]. Although the references, given above, contain excellent and detailed accounts of the theories, it is worth while discussing their assumptions, successes and limitations in order to put them into perspective with current thoughts on oxidation catalysis.

A. Multiplet Theory

The multiplet theory of catalysis supposes that only part of a molecule undergoing catalytic conversion actually participates in the reaction. This particular group of atoms is termed the "index group" and may be considered analogous to the functional groups of organic chemistry. The theory further assumes that only certain atoms of the catalyst possess the necessary configuration for the reaction; these atoms are termed the "multiplet". According to the theory, during catalysis, the index group of a reactant molecule is superimposed on the active atoms of the catalyst, thereby yielding an intermediate multiplet complex within which bond deformation and migration occurs.

To illustrate these concepts, consider the reaction

$$AB + CD \rightarrow AD + BC$$

126

From the standpoint of the multiplet theory, this reaction may be written:

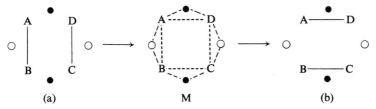

(a)   M   (b)

In this scheme, A, B, C and D represent reacting atoms such as C, O, N etc., which may contain substituents, and $\circ \bullet \circ$ represents the multiplet. The usual way to represent a reaction, according to the multiplet theory, is to show only the atoms and bonds of the index group in such a way that, during reaction, two vertical bonds are broken and two horizontal bonds are formed. For example, the dehydrogenation of isopropanol is classified using the index:

Transition from state (a) to state (b) is accomplished via the unstable complex M, which, depending on thermodynamic considerations, either decomposes giving (b) or reverts to (a).

Considerable evidence exists to support the idea of the participation of similarly oriented index groups in certain types of reaction. Thus, in an investigation of the kinetics of the dehydrogenation of amines to ketimines $(R_1R_2CHNH_2 \rightarrow R_1R_2C=NH + H_2)$, Balandin and Vasyunina [689] found that reaction took place in the presence of several catalysts, including palladium, alumina-supported nickel and thoria. A number of amines were used and it was found that the individual activation energies for dehydrogenation remained remarkably constant for the same catalyst. For example, activation energies of 9·7, 9·9, 9·1 and 11·4 kcal. mole^{-1} (Pd) and 9·4, 9·6, 8·7 and 10·7 kcal. mole^{-1} (Ni) were obtained for 2-aminoheptane, 2-methyl-4-aminopentane, 2,4-dimethyl-3-aminopentane and 1-diethyl-4-aminopentane respectively. It seems very reasonable, therefore, to classify this particular reaction:

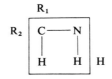

As a further example, Balandin and Isagulyants [690] observed that some substituted cyclohexanes were dehydrogenated, via cyclohexenes, to yield substituted benzenes with very similar activation energies. On Cr_2O_3, for example, the activation energies for cyclohexane, methylcyclohexane and 1,3-dimethylcyclohexane were 25·9, 23·7, and 22·2 kcal. mole^{-1} respectively which apparently suggested an edgewise orientation of the ring to the catalyst. This may be depicted:

This particular mode of cycloalkane dehydrogenation is termed a doublet mechanism.

On account of the index groups, certain restrictions, with respect to the catalyst, are imposed on one's choice of catalyst by virtue of the stereochemistry of a reaction and the interatomic distances within the index group. This may be illustrated by the so-called sextet model of cyclohexane dehydrogenation to benzene which was proposed for the reaction occurring at about 300°C in the presence of some metals. Since benzene is a flat molecule, the sextet model envisages the flattening of the reactant ring during the reaction. A six-membered carbon ring may be flatly superimposed, e.g. on the (111) facet of certain metals in which the atoms are arranged in equilateral triangles. According to the sextet model, flattening occurs on formation of the multiplet complex, the forces promoting chemisorption being stronger than those hindering changes in the conformation of the reactant molecule. The model also anticipates that only metals possessing a face-centred cubic or hexagonal lattice will be active catalysts for this particular reaction, since only with these structures is the equilateral triangular arrangement of atoms encountered. Thus, the alloy 74·9% Co–21·7% Fe (fcc lattice) is active, whilst the alloys 50% Co–50% Fe and 24·2% Co–75·8% Fe, which crystallize as body-centred cubes, are inactive [686].

Other examples to support the "principle of structural correspondence" within the multiplet theory have been discussed by Balandin [686] but two cases are particularly notable and involve (i) olefin hydrogenation and (ii) the dehydration-dehydrogenation of alcohols. In the former case (index

),

an interesting rule, involving preservation of the valence angle, was discovered.

During the chemisorption of, say, ethylene on two atoms (K) of a catalyst, the double bond was assumed to open, yielding the configuration

In this state, θ must be close to the tetrahedral angle ($\sim109°$), and, since distances C–C and C–K could be calculated from the sum of the atomic radii, an estimation of the optimum interatomic distance K–K was possible. Several metals, including Pt and Fe, possess structures with the correct spacing and, indeed, such metals proved to be catalytically active towards olefin hydrogenation. In the second case, Balandin [686] discussed the investigations of Rubinstein and Pribytkova [691] on the catalytic reactions of alcohols in the presence of magnesia. Depending on the distance (a Å) between the atoms of the catalyst, alcohols undergo either dehydration or dehydrogenation over magnesium oxide. Rubinstein found that, depending on the method of preparation, MgO could be obtained having different lattice parameters, a. It was found that the greater the value of a, the more dehydration was favoured. Considering the parameters for the index group involved (see Fig. 4), Rubinstein's results were easily explained on the basis of the multiplet theory.

Despite such successes, the principle of structural correspondence (i.e. the geometrical factor) was insufficient to explain the products of certain catalytic reactions. For example, Tolstopyatova [692] found that ThO_2 and CaF_2, although possessing identical structures and almost the same interatomic distances, behaved quite differently towards the catalytic decomposition of alcohol. Dehydration was the preferred reaction in the presence of thoria whilst preferential dehydrogenation occurred over calcium fluoride. This observation indicated that alcohol molecules were differently orientated with respect to the surface of the catalyst. According to Balandin [686], the reason for this difference lay in the energetics of the reaction. Thus, in addition to structural correspondence, the multiplet theory also had to postulate energetic correspondence.

According to the multiplet theory, the most important energy factors involved in heterogeneous catalysis include the heat of reaction (ΔH) and the activation energy (E_a); ΔH and E_a may be derived from fundamental considerations of bond energies.

Consider the energy required to break the A–B bond in a molecule AB during adsorption on a catalyst. If the bond dissociation energy A–B of AB in the gas-phase is denoted by Q_{A-B} and those between the individual atoms A and

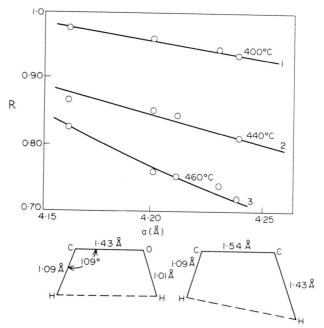

FIG. 4. The dependence of the relationship R, of the percentage dehydrogenation to the percentage dehydration of n-Butanol on the lattice parameter $a(R)$ of magnesium oxide (Fig. 13, p. 39. Balandin, *Adv. Catal.* **19** (1969)).

B and the catalyst surface (K), i.e. A–K and B–K, are denoted by Q_{A-K} and Q_{B-K}, then:

$$A\text{–}B + 2K \rightarrow A\text{–}[K] + B\text{–}[K] - E_{AB, K} \tag{1}$$

where $E_{AB, K}$ is the energy required to break bond A–B on adsorption ($\neq Q_{AB}$) and A–[K] and B-[K] represent adsorbed atoms of A and B. Other relevant equations include:

$$A\text{–}B \rightarrow A + B - Q_{AB} \tag{2}$$
$$A + K \rightarrow A\text{–}[K] + Q_{AK} \tag{3}$$
$$B + K \rightarrow B\text{–}[K] + Q_{BK} \tag{4}$$

whence

$$A\text{–}B + 2K \rightarrow A\text{–}[K] + B\text{–}[K] - Q_{AB} + Q_{AK} + Q_{BK} \tag{5}$$

and

$$E_{AB, K} = -Q_{AB} + Q_{AK} + Q_{BK} \tag{6}$$

If, in comparison to the gas-phase

$$Q_{AB} > Q_{AK} + Q_{BK} \tag{7}$$

then bond A–B will only be weakened on adsorption. If, however,

$$Q_{AB} < Q_{AK} + Q_{BK} \tag{8}$$

the bond A–B will be completely broken and structural factors will not influence the reaction.

In the case of the doublet reaction

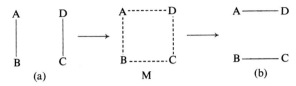

two bonds, A–B and C–D, are either broken or stretched. The heat of formation of the multiplet complex, M, is then given by

$$E = E_{AB, K} + E_{CD, K} \tag{9}$$

whilst the energy required for the further decomposition of M into products, is given by the expression:

$$E = E_{AD, K} + E_{BC, K} \tag{10}$$

Now, the energy required for the formation of M from the initial reactants is given by the equation:

$$E' = -Q_{AB} - Q_{CD} + (Q_{AK} + Q_{BK} + Q_{CK} + Q_{DK}) \tag{11}$$

and the energy required for the decomposition of M into products may be written:

$$E'' = Q_{AD} + Q_{BC} - (Q_{AK} + Q_{BK} + Q_{CK} + Q_{DK}) \tag{12}$$

if $-E' < -E''$, M will decompose into the initial reactants, but if $-E'' < -E'$, the reaction will proceed to state (b).

Quantitatively, Balandin derived a relationship between $E(E = \min(E', E'', L)$ where L is a constant, generally $= 0$ or A) and the activation energy (E_a) of the reaction [686]. This relationship is expressed by the equation

$$E_a = A - \gamma E \tag{13}$$

where, for endothermic reactions, $\gamma = 0.75$ and $A = 0$; whilst for exothermic reactions, $\gamma = 0.25$ and $A \simeq 11.5$ kcal. Thus, from bond energy considerations the activation energy of a reaction may be obtained from the multiplet theory.

Balandin [693] also considered the case in which bond energies, Q, were regarded as variable. Under these conditions, certain other relationships were introduced, viz:

$$-u = Q_{AB} + Q_{CD} - Q_{AD} - Q_{BC} \tag{14}$$

$$s = Q_{AB} + Q_{CD} + Q_{AD} + Q_{BC} \tag{15}$$

$$q = Q_{AK} + Q_{BK} + Q_{CK} + Q_{DK} \tag{16}$$

where

$u =$ heat of reaction

$s =$ sum of the energies involved in breaking and forming bonds and

q = adsorption potential of the catalyst.

Substituting equations (14)–(16) into equations (11) and (12), the following expressions were obtained:

$$E' = q - \tfrac{1}{2}s + \tfrac{1}{2}u \tag{17}$$

and

$$E'' = -q + \tfrac{1}{2}s + \tfrac{1}{2}u. \tag{18}$$

Plotting values of E against q, equations (17) and (18) yield intersecting straight lines (volcano-shaped curves). The coordinates of the point of intersection were found to have values

$$q_0 = \tfrac{1}{2}s \tag{19}$$

and

$$E_0 = \tfrac{1}{2}u. \tag{20}$$

Equation (19) was said to express the principle of energetic correspondence for catalytic doublet reactions. Thus, for the selection of an active catalyst for an endothermic reaction, it was stated that it is necessary for the adsorption potential, q, of the catalyst to be as near as possible to $\tfrac{1}{2}s$, i.e. the average of the energies of broken- and newly-formed bonds.

In his review, Balandin [686] gave detailed methods for obtaining values of Q_{AK} etc. for use in such calculations. It is beyond the scope of this book to discuss the implications of the findings. Suffice it to say that application of equations (9) etc. to the prediction of the course of many catalytic reactions, met with remarkable success, the theory predicting, often precisely, the experimental products found in reactions such as hydrogenolysis, ring-opening, dehydrogenation of hydrocarbons etc.

Balandin's theory has its critics, however. For example, Clark [694], although admitting the considerable evidence in support of the multiplet theory, points out that there is much evidence which casts doubt on the validity of correlations between bulk lattice parameters and catalysis. Most important, in Clark's opinion, are "the objections raised by the results of LEED, (and) electron microscopy . . . which in some cases reveal considerable disruption of the surface, so that bulk lattice parameters have little meaning with respect to surface geometry". It is the present author's opinion, however, that such drawbacks become insignificant when compared with the order which Balandin's concepts imposed on the then-chaotic state of catalysis.

B. Electronic Theory

The electronic theory of catalysis on semiconductors described by Wolkenstein [685] approaches the problems of catalysis from a standpoint totally

different from that used in the multiplet theory. Emphasizing that every catalytic process begins with the act of adsorption, it proposes that any generalized
theory of catalysis should proceed from a consideration of theory of adsorption.

Having as a basis the bond theory of solids, which explained the electrical
properties of metals and semiconductors in terms of the formation of energy
bands from the original atomic orbitals of the individual atoms, the electronic
theory regards particles adsorbed on the surface of a catalyst as impurities or
structural defects disturbing the strictly periodic structure of the surface. The
concept of a chemisorbed particle as a structural defect allows one to regard the
particle as a centre of localization of either a free electron (an acceptor) or a
free hole (a donor). Free electrons and holes in a crystal may arise in a number
of ways. For example, the presence of Zn^+ in the array of Zn^{2+} ions in zinc
oxide, constitutes a free electron, whilst a free hole may be regarded as an O^-
ion amongst the O^{2-} ions.

The localization of a free electron or hole near a chemisorbed particle, means
that there is the possibility of their participation in the bonding of the particle.
Even so, chemisorption may be either "weak" or "strong". Thus in "weak"
bonding (denoted as CL by Wolkenstein [685] where L = symbol of lattice and
C = chemisorbed particle), the particle interacts with neither a hole nor an
electron, remaining electrically neutral. When a free electron or hole participates directly in the bond, strong n- or acceptor bonds (CeL, where eL = free
electron of the lattice) or strong p- or donor bonds (CpL, where pL = free hole)
arise.

Since free electrons and holes may participate significantly in chemisorption,
they are important agents in catalysis and may be regarded as free (unsaturated)
valencies wandering through a crystal. According to Wolkenstein [685], free
valencies possess the following properties:

(1) They can appear and disappear continuously.
(2) They migrate through a lattice so that, in a perfect crystal, there is an equal
    probability of finding a free valence at any point.
(3) Their equilibrium concentration depends, not only on the nature of the
    crystal, but also on external factors such as temperature, radiation, etc.
(4) They are continuously exchanging with the bulk of the crystal, which acts
    as a reservoir of free-valencies.

The participation of surface free-valencies in chemisorption may lead to the
chemisorbed species possessing enhanced reactivity, i.e. chemisorption may
lead to the transformation of a valence-saturated particle into an ion-radical
and vice-versa. The diagram below illustrates "weak" and "strong" acceptor

bonding for the oxygen molecule

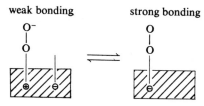

The intervention of free valencies in chemisorption may also cause the dissociation of a molecule on adsorption. For example, consider a molecule AB, comprising two monovalent positive ions (e.g. H_2), approaching the surface of a semiconductor

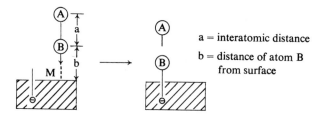

a = interatomic distance

b = distance of atom B from surface

As the molecule approaches the surface, the free lattice-electron becomes more and more localized near the point of approach (M in the diagram). As atom B approaches M, the bond between B and the surface grows stronger, whilst the bond B–A grows weaker. Eventually, as distance b decreases and a increases, a point is reached where B becomes attached to the surface by a strong *n*-bond, whilst A is either freed or weakly-attached to the surface. The overall reaction may be represented by the equation:

$$AB + eL \rightarrow ABeL \rightarrow A + BeL$$

The dissociation of a molecule on adsorption may also proceed in such a manner that both dissociation products form "weakly" chemisorbed species. Such a case may be encountered with the oxygen molecule; the double bond between the atoms may be broken as the result of the transfer of two electrons from two negative ions of the lattice to the molecule, thus forming two localized holes:

The oxygen atoms thus formed are attached to the surface by "weak" bonds and may be considered as either inactive or only slightly reactive. The subsequent delocalization of the hole, or the free-electron which combines with the hole, then brings the oxygen atom into a reactive state.

The electronic theory of catalysis regards chemisorbed species as inter-convertible, i.e. a species may pass from a state having one type of bonding into a state with a different type. Such transitions may be best illustrated by considering a diagram of the energy bond scheme of a semiconductor (Fig. 5).

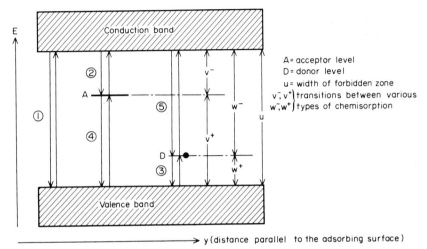

FIG. 5. Energy band scheme of a semiconductor.

A foreign particle "weakly" chemisorbed on a surface has a counterpart in the energy-band scheme of a crystal. A particle possessing an affinity for a free electron is depicted by an acceptor level (A); a particle possessing affinity for a hole corresponds to a donor level (D). When a particle has an affinity for both, it is depicted by the simultaneous presence of A and D. The positions of A and D in the forbidden zone depend on the nature of the lattice and the adsorbed particle C.

Electron transitions can take place between the valence band and the conduction band (transition 1) as well as between the energy bands and the local levels (transitions 2–5). The appearance of an electron on the level A represents the transition of a chemisorbed particle C from a state of "weak" to a state of "strong" acceptor bonding with the surface. This may be effected in two ways:

(1) By a free electron of the conduction band falling on level A.
(2) By an electron of the valence band being promoted to A.

The removal of an electron from level D represents the transition from "weak" to "strong" acceptor bonding. Again there are two ways to bring this about:

(1) By the recombination of an electron in D with a free hole wandering in the valence band.
(2) By the ejection of an electron from D to the conduction band.

These transitions may be written in the following way:

	Energy	Transition
$eL + pL \rightleftharpoons L$	u	1
$CL + eL \rightleftharpoons CeL$	v^-	2
$CL + pL \rightleftharpoons CpL$	w^+	3
$CeL + pL \rightleftharpoons CL$	v^+	4
$CpL + eL \rightleftharpoons CL$	w^-	5

In these equations, a partial arrow from left to right denotes an exothermal transition whilst one from right to left depicts an endothermal transition.

From the foregoing account it may be seen that semiconductors may catalytically influence either radical reactions or donor–acceptor reactions. In the former case, the participation of free electrons or holes, on the catalyst, in chemisorption, results in the chemisorbed particles spending part of their time on the surface in a radical state, leading to an increase in their reactivity. The role of the catalyst is to produce surface radicals which appear at the expense of free valencies of the catalyst existing on the surface or appear there in the course of reaction.

Not all particles chemisorbed on the surface of a semiconductor are reactive, however, and Wolkenstein [685] showed, using Fermi statistics, that the principal chemisorptive and catalytic properties of a catalyst are determined by the position of the Fermi level on the surface. The concept of the Fermi level arises from the free-electron theory of metals wherein Fermi statistics showed that the probability of finding an electron in a given energy level is given by the expression:

$$f = \{\exp{(E - E_{max})}/kT + 1\}^{-1}$$

where E_{max} = Fermi level.

At absolute zero, when $E > E_{max}, f = 0$; when $E < E_{max}, f = 1$, At any temperature, therefore, where $E = E_{max}, f = \frac{1}{2}$.

The Fermi level, it was shown [685], determines the following:

(1) The total number of chemisorbed particles on a surface in equilibrium with the gas phase at a given temperature and pressure.
(2) The magnitude and sign of the surface charge for a given coverage.
(3) The relative amount of the different forms of chemisorption, on the surface, which are distinguished by the character of the bonds between the chemisorbed particle and the surface.
(4) The reactivity of the chemisorbed particles, i.e. the probability of a particle being in a radical- or a valence-saturated state.
(5) The catalytic activity of the surface for a given reaction.
(6) The catalytic selectivity for two (or more) simultaneous reactions.

Since the position of the Fermi level uniquely determines the concentration of the electron- and hole-gases on the surface of the crystal, it is of paramount importance to elucidate the factors which determine the position of the Fermi level on the surface of the crystal. To do this, Wolkenstein [685] considered, initially, the consequence of the existence of a "strong" form of chemisorption, i.e. the form wherein the chemisorbed particle attaches itself to a free hole or electron of the crystal lattice. On account of this, a charge appears on the surface layer of the semiconductor; this induces a space charge at the surface, opposite in sign and magnitude, and consequently deformation of the energy bands near the surface of the semiconductor. This is shown in Fig. 6. In this diagram, (a)

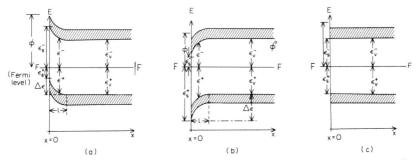

FIG. 6. Bending of the energy bands at the surface of a semiconductor according to Wolkenstein (Ref. 685).

corresponds to a negative surface charge, (b) to a positive surface charge and (c) to an electrically neutral surface. The x-axis is directed into the crystal, perpendicular to the adsorbing surface which coincides with the surface at $x = 0$. The distance, 1, in (a) and (b) is called the "screening length" and corresponds to the distance at which bending of the bands may be considered significant compared with kT. ϕ is called the thermoelectric or thermodynamic work function.

Once electronic equilibrium has been established, the surface and the bulk of the semiconductor have a common Fermi level, but, due to bending of the bands, the position of the Fermi level in the energy spectrum of the crystal depends on the distance from the surface and is characterized by its distance from the top of the valence band (ε^+). The position of the Fermi level can be described by the expressions:

$$\varepsilon_{x=0} = \varepsilon_s^+$$
$$\varepsilon_{x=\infty}^+ = \varepsilon_v^+$$
$$\varepsilon_{x>1}^+ = \varepsilon^+$$

The reaction rate for a given surface coverage depends on the relative amount of the active forms of chemisorbed molecules and this is further related to the position of the Fermi level in the energy spectrum of the crystal. For instance, the relative coverage of particles on a surface, found in states of "weak" (η^0), "strong" acceptor (η^-) and "strong" donor (η^+) bonding, are described by the formulae

$$\eta^0 = \frac{N^0}{N}; \qquad \eta^- = \frac{N^-}{N}; \qquad \eta^+ = \frac{N^+}{N}$$

where $\eta^0 + \eta^+ + \eta^- = 1$, N^0, N^- etc. are equal to the number of particles of a given type chemisorbed on unit surface and N is equal to the total number of particles adsorbed on unit surface. According to Fermi statistics:

$$\frac{N^-}{N^0 + N^-} = \frac{1}{1 + \exp[(E_s^- - v^-)/kT]} \quad \text{and} \quad \frac{N^+}{N^0 + N^-} = \frac{1}{1 + \exp[(E_s^+ - w^+)/kT]}$$

where E_s^- and E_s^+ are respective distances from the Fermi level.

Taking as an example the catalytic decomposition of ethanol, Wolkenstein [685] discussed the effect, on the selectivity of a process, of the position of the Fermi level in a catalyst. The alternative paths considered for the decomposition of ethanol were dehydrogenation to yield acetaldehyde + hydrogen and dehydration to give ethylene and water. The mechanisms for these reactions, given by Wolkenstein [685], are as follows:

(i) *Dehydrogenation*

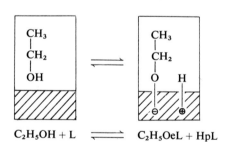

Step 1 Adsorption $C_2H_5OH + L \rightleftharpoons C_2H_5OeL + HpL$

Step 2 Surface reaction $C_2H_5OeL + \overset{\cdot}{H}L \longrightarrow \overset{\cdot}{C_2H_4OeL} + H_2$

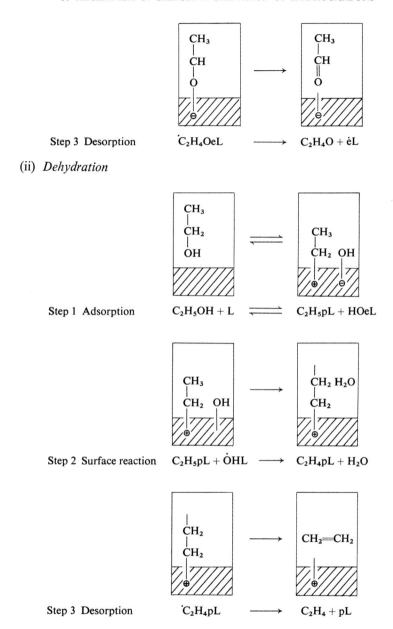

Step 3 Desorption $\dot{C}_2H_4OeL$ $\longrightarrow$ $C_2H_4O + \dot{e}L$

(ii) *Dehydration*

Step 1 Adsorption $C_2H_5OH + L$ $\rightleftharpoons$ $C_2H_5pL + HOeL$

Step 2 Surface reaction $C_2H_5pL + \dot{O}HL$ $\longrightarrow$ $C_2H_4pL + H_2O$

Step 3 Desorption $\dot{C}_2H_4pL$ $\longrightarrow$ $C_2H_4 + pL$

In these schemes, the reaction path is determined by the first step and depends on which bond is broken on adsorption. This in turn is dictated by the nature of the catalyst. Both types of decomposition may occur on the same catalyst, the relative activity depending on the position of the Fermi level. A lowering of the Fermi level, for example, favours dehydration. Factors which lower the Fermi level (e.g. an acceptor level), poison the catalyst towards dehydrogenation and promote dehydration, and vice-versa.

Reactions which proceed faster as the Fermi level is raised are accelerated by electrons and are known as n-type or acceptor reactions; opposite reactions are known as p-type or donor reactions.

Not only different reactions, but also different stages of a heterogeneous reaction may belong to different classes. As an example, Wolkenstein [685] considered the oxidation of carbon monoxide which may be catalysed by either n-type (e.g. ZnO) or p-type (e.g. NiO) semiconductors. Considering the possible mechanisms of this reaction, it was initially assumed that the surface was covered with chemisorbed atomic oxygen which, in the ion-radical state, served as adsorption centres for carbon monoxide

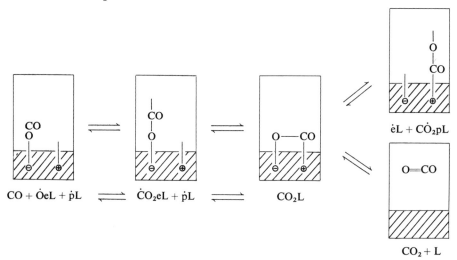

Different mechanisms therefore operate when (a) the reaction is limited by the adsorption of carbon monoxide and (b) the reaction is limited by desorption of carbon dioxide.

An important consequence of the electronic theory is the relationship between the catalytic activity and the electrical conductivity of a catalyst. The latter factor is uniquely determined by the position of the Fermi level inside the crystal (ε_v^+) and the degree of bending of the bands. The relationship between the electron- and hole-components of conductivity depends on the position of the Fermi-level within the forbidden zone. The higher the Fermi level (i.e. the closer to the conduction band), the bigger will be the electron component and the less significant will be the hole component. If one of these components may be neglected in comparison with the other, then n-type or p-type semiconductors arise. When both components are equal in magnitude, "intrinsic" or "mixed" semiconductivity emerges.

Catalytic activity is determined by the position of the Fermi level on the surface of the catalyst (ε_s^+). This, in turn, is determined by its position within the catalyst (ε_v^+). As ε_v^+ increases, so ε_s^+ increases, and vice-versa. Factors altering ε_v^+ (and hence electrical conductivity) will also alter ε_s^+ (and hence the catalytic activity). This gives rise to the parallelism between changes in electrical conductivity and catalytic activity often observed in catalysis.

error

A notable success of the electronic theory has been to offer an explanation for the mechanism of promotion or poisoning by apparently insignificant amounts of impurities added to semiconductors. It was shown that ε_s^+ can be varied to some extent by doping the bulk of a crystal with the appropriate impurity. Thus the impurity regulates the surface concentrations of the electron and hole gases which regulate the reaction rate. Impurities need not necessarily be chemically foreign atoms, but include any local disturbances in the strictly periodic nature of the lattice, such as vacancies and atoms of the lattice which have been thrown into interstitial positions or onto the surface of the crystal. There are two types of impurity—donor and acceptor—and any doping of a crystal usually leads to displacement of the Fermi level on the surface. Surface impurities directly affect ε_s^+ leaving ε_v^+ unchanged; bulk impurities alter ε_v^+ and hence ε_s^+. Acceptor impurities lower the Fermi level whilst donor impurities raise the level.

Despite the fact that the electronic theory of catalysis, documented by Wolkenstein [685], is capable of explaining certain empirical correlations, such as the covariance of catalytic activity and electrical conductivity or work function, other factors existed which could not be easily explained. Amongst these may be mentioned:

(1) The existence of correlation between the catalytic activity of a semiconductor and the width of its forbidden zone (u), the activity increasing as u falls.

(2) The fact that, for a series of catalyst samples prepared from a base semiconductor by doping with donors (or acceptors), the catalytic activity of the samples does not always mirror its electron- (or hole-)concentration as one would expect according to the electron theory. However, recent work by Lee and Mason [695] and Lee [696] satisfactorily explains these experimental findings in terms of a model of catalysis which contains the postulate that the energy necessary for surface reaction on a semiconductor may be provided by the production of electron-hole pairs and their annihilation during the reaction.

C. Crystal and Ligand Field Models

During a study of H_2–D_2 exchange on certain oxides of the fourth period of the periodic table, Dowden and Wells [687] observed that activity was almost zero with TiO_2 (d-electronic configuration of Ti^{4+} is d^0) but rose to a maximum at Cr_2O_3 (d^3) before falling to a minimum again at Fe_2O_3 (d^5). An even higher peak of activity was found with Co_3O_4 (d^{6-7}) but this fell again rapidly at CuO and ZnO (d^9 and d^{10}). Measurement of the catalytic activity of fourth-period oxides for other reactions, e.g. the dehydrogenation and disproportionation of cyclohexane [697] and the hydrogenation of ethylene [698], also showed two peaks, one at Cr_2O_3 and the other at Co_3O_4.

In the opinion of Dowden and Wells [687], simple electronic theory was unable to explain the changes of catalytic activity when proceeding across the fourth period of the periodic table, since the electronic levels of their metal cations (referred to vacuum) are approximately equal. According to Dowden and Wells [687], however, a possible explanation lies in the fact that chemisorption on a surface transition metal cation increases the number of ligands on the cation, thus producing a change in its coordination with a resultant increase in symmetry and a closer approximation to the situation of a metal ion in the bulk of the lattice. For example, during the adsorption on the (100) face of a NaCl-type lattice, the number of ligands attached to a cation will change from five to six and the shape of the coordination sphere from a tetragonal pyramid to an octahedron:

$$L^{n-} \diagdown \quad \diagup L^{n-} \atop \quad M^{m+} \quad + Y_2(g) \quad \longrightarrow \quad L^{n-} \diagup \quad \diagdown L^{n-}$$

On a (100) face of a similar lattice, adsorption will lead to the changes tetrahedron → tetragonal pyramid → octahedron.

Changes in coordination, such as these, bring about a change in the crystal field stabilization energy of the ion and Dowden and Wells [687] have calculated these changes for all changes of coordination such as the ones given above. In the majority of cases, an increase in the number of ligands around a cation resulted in an increase in the stabilization energy, except for cations with the electronic configuration d^0, d^5 and d^{10} which, in weak crystal fields, have no additional stabilization. The twin peaks of activity, previously mentioned, are explicable if the rate-determining step of the reactions involved an increase in the coordination number of the ion by adsorption or some other step. In that case, the minimum effect would be observed with d^0, d^5 and d^{10} systems, whilst the most catalytically active systems would have the configurations d^3, d^6 and d^8, since these show the maximum changes in crystal-field stabilization energy during an increase in coordination number.

Although the crystal-field theory, used in conjunction with the electronic theory of catalysis on semiconductors, has been very useful in predicting the best catalyst for certain reactions (e.g. H_2–D_2 exchange, decomposition of N_2O), a number of critics have expressed doubts about the data on which the theory is based. These arguments have been thoroughly reviewed by Krylov [699], who further points out that in "the overwhelming majority of recent works, the two-spiked diagram for change of catalytic activity of metal oxides of the fourth period in oxidation-reduction reactions is verified". A significant failure of Dowden's theory might perhaps be thought to have been found in results such as those of Shelef *et al.* [700], but, quoting Krylov [699] once more, "the non-occurrence of this relationship (i.e. twin peaks) in a given case does not justify considerations of arguments against the expedience of applying crystal field theory in catalysis".

6.2. Correlations Within Oxidation Systems

Perhaps the greatest disadvantage of the "universal" theories of catalysis lies in their attempts to correlate too many reactions and too many catalysts. As Roginskii [701] has pointed out, the diversity of catalysis and the mechanisms involved preclude the possibility of finding a comprehensive solution. More effective correlations will probably emerge only when the classes of reactions are restricted. For this reason, subsequent sections of this chapter will be confined to systems involving oxygen as a reactant in one form or another.

A. Isotopic Exchange of Oxygen

Probably the simplest reaction occurring with the involvement of oxygen, is the isotopic exchange reaction:

$$^{16}O_2 + {}^{18}O_2 \rightleftharpoons 2\,{}^{16}O^{18}O$$

Isotopic exchange of oxygen has recently been reviewed by Novakova [702] and Parravano [703]. Novakova, for instance, discusses reactions involving not only $^{18}O_2$, but also ^{18}O-labelled water, carbon monoxide and carbon dioxide. The phenomenon was first studied by Winter [704] but it has been exploited considerably by Boreskov and co-workers [705–708].

According to Boreskov [705], two types of oxygen exchange can occur: (i) homomolecular (involving unpairing the electrons in the oxygen–oxygen bond of the oxygen molecule); and (ii) heterogeneous exchange (electrons not unpaired). Homomolecular exchange can occur either by a dissociation-association mechanism or by the formation of three- or four-atom complexes on a catalyst. The activity of various catalysts, e.g. transition metal oxides, towards homomolecular exchange apparently permits one to evaluate the probability of the formation of intermediate forms of oxygen on the catalyst surface [705]. Heterogeneous exchange of oxygen with the oxygen of a solid lattice allows one to evaluate both the reactivity of this oxygen and its distribution. In the opinion of Boreskov [705], most information is obtained by the simultaneous study of homomolecular and heterogeneous exchange.

Boreskov [705] has found that, in the case of oxides, the rate of both homomolecular and heterogeneous exchange of oxygen depends markedly on the pre-history of the oxide. For numerous oxides there appear to be two types of surface, differing considerably in their activity towards isotopic exchange. One type of surface state may be obtained by the high-temperature treatment of the oxides in a stream of gaseous oxygen and corresponds to a surface layer containing an equilibrium concentration of oxygen. In this state, the surface possesses a stable and reproducible activity towards homomolecular exchange which takes place at fairly high temperatures. The other type of surface state may be achieved after high-temperature treatment of the oxides in a vacuum. This is characterized by a high rate of homomolecular exchange, even at very low temperatures. This activity does not remain steady, however, and may be eliminated by heating the oxide in oxygen.

When an equilibrium concentration of oxygen is present in a surface, the rates of homomolecular and heterogeneous exchange are often very close [705]. This suggests that, on such oxides, the type of intermediate interaction between the oxygen and the catalyst is either identical to that existing in the actual surface layer of the oxide or so close that rapid transitions between the two may occur. For oxides with uniformly surface-bonded oxygen, all the oxygen of the oxide participates although in cases where differently-bound forms exist, only the most active oxygen participates, i.e. catalytic activity towards the homomolecular exchange of oxygen may be used as a measure of the bonding energy and reactivity of surface oxygen in equilibrated oxide catalysts.

As has been said, oxides prepared in a vacuum behave very differently, displaying only homomolecular exchange. In other words, lattice oxygen does not participate in the reaction. The exchange reaction may take place via a dissociation-association mechanism yielding oxygen adatoms which differ from lattice oxygen ions, or by the formation and decomposition of three- or four-atom complexes. Unfortunately, active sites, formed as a result of the high-temperature treatment of oxides in vacuo, are not sufficiently stable to establish their participation in any reaction other than the isotopic exchange of oxygen.

Finally, returning to oxides containing the equilibrium content of oxygen in the surface layer, close correlations have been found between isotopic exchange reactions and a number of other reactions including the oxidation of hydrogen, carbon monoxide and butenes [706, 708]. This correlation exists, not only for single oxides, but also for complex oxides such as cobalt spinels [709]. This correlation is understandable if it is assumed that, for certain groups of catalysts, the variation in the energy of the activated complex on the surface is determined essentially, by the variation in the energy of one of the bonds being either broken or formed. If the formation of an active complex in an oxidation reaction involves breaking the oxygen–catalyst bond, then it is to be expected that there will be a linear dependence of the activation energy on variations in bonding energy.

In conclusion, Haber and Grzybowska [710] have recently complied a table of oxygen exchange data, found in the literature, for certain simple oxides and also some selective oxidation catalysis (Table 36):

Table 36. Activity of some oxides towards isotopic oxygen exchange

Catalyst	T(°C)	Rate of exchange (g $O_2/m^2/h$)	Ref.
MoO_3	580–601	9×10^{-4}	[217]
Bi/Mo = 2 : 1	250–500	None	[216]
Bi/Mo = 1 : 1	474–500	None	[198]
Co/Mo = 1 : 1·7	599–634	$1·8 \times 10^{-4}$	[711]
Co/Mo = 2 : 1	401–462	$2·9 \times 10^{-2}$	[711]
Co_3O_4	125–250	12·7	[712]
Fe/Mo = 1 : 1·7	511–551	7×10^{-3}	[711]
Fe/Mo = 1 : 1	508–552	1×10^{-3}	[711]
Fe_2O_3	350–450	4×10^{-1}	[702]

From this, it appears that highly selective oxidation catalysts, such as bismuth molybdate, show little activity towards oxygen exchange. Fairly selective catalysts, such as molybdenum(VI) oxide and Co–Mo oxides, show a very low rate of exchange and catalysts such as Co_3O_4, which are effective for the complete oxidation of hydrocarbons, have a high rate of exchange although, apparently, the amount of oxygen exchanged is less than a monolayer.

B. Complete Oxidation of Hydrocarbons

Since the report by Fahrenfort *et al.* [713] that the activity of certain metals towards the decomposition of formic acid could be correlated with the heat of formation of the formates of those metals, thermodynamic quantities such as the heats of formation of certain compounds have been extensively used in attempts to discover unifying trends in catalytic systems. Typical of such studies is the work of Moro-oka *et al.* [714, 640].

For instance, Moro-oka and Ozaki [714] studied the oxidation of propylene over a large number of metal oxides using hydrocarbon-oxygen mixtures above (series H) and below (series L) the explosive region for propylene-oxygen. Carbon dioxide was the sole oxidation product for both series of experiments, although, in the case of vanadium, thorium and aluminium oxides, some carbon monoxide was also formed. Moro-oka and Ozaki [714] determined a number of kinetic parameters, such as reaction rate at 300°C, reaction orders and Arrhenius parameters, for the oxidations and found that the activity of the oxides could be correlated with a parameter (ΔH_0) representing the heat of formation of the oxide divided by the number of oxygen atoms in the oxide molecule. It was found that, by plotting the catalytic activity against ΔH_0, a fairly smooth curve was obtained and the larger the value of ΔH_0, the less active was the catalyst and the higher the reaction order in propylene. Moro-oka and Ozaki [714] regarded ΔH_0 as an indication of the metal–oxygen bond strength in the oxides; unfortunately, however, the oxidation state of the metal ion in the oxides was never precisely known but "judged from the colour of the catalyst".

Moro-oka *et al.* [640] continued their investigations into the catalytic properties of various oxides with respect to the total oxidation of hydrocarbons by studying the reactions of isobutene, acetylene, ethylene and propane. Again it was established that a relationship existed between ΔH_0 and the catalytic activity of an oxide. In the case of isobutene and acetylene, a relationship was also discovered between the reaction order in hydrocarbon and oxygen and ΔH_0; as ΔH_0 increased, so the reaction order in hydrocarbon increased whilst the order with respect to oxygen decreased. This variation was explained on the basis of the adsorption strength of the catalyst towards the hydrocarbon and, from results obtained on the competitive oxidation of certain pairs of hydrocarbons, it was concluded that hydrocarbons reacted via the adsorbed state, the sequence of adsorption strength being:

isobutene > acetylene > propylene > ethylene > propane.

Moro-oka *et al.* [640] then went on to classify their catalysts according to the following criteria:

(1) Catalysts having a low value of ΔH_0 (e.g. Pd, Pt; $\Delta H_0 = 16\cdot0$ and $20\cdot4$ kcal. mole^{-1}/O atom, respectively); reaction is characterized by having a negative order in hydrocarbon and an order with respect to oxygen of almost one. This suggested to Moro-oka *et al.* [640] that, with oxides of this type, the surface was covered with hydrocarbon and the slow step of oxidation was chemisorption of oxygen on this surface.

(2) Catalysts having medium values of ΔH_0 (e.g. Co$_3$O$_4$, Fe$_2$O$_3$; $\Delta H = 48\cdot3$ and $67\cdot6$ kcal. mole^{-1}/O-atom, respectively); reaction is characterized by having reaction orders, with respect to hydrocarbon, which are approximately zero or have low positive values and a reaction order with respect to oxygen of *ca.* $0\cdot5$.

(3) Catalysts having high values of ΔH_0 (e.g. $CeO_2 = 122\cdot7$ kcal. $mole^{-1}/O$ atom). Reaction in the presence of such catalysts are first-order in hydrocarbon and zero-order in oxygen. It was suggested that, with these catalysts, the surface was almost completely covered with oxygen and the slow step involved either chemisorption or reaction of the hydrocarbon on the oxygen-covered surface.

Even within the author's own results [714], there are a number of exceptions to this classification, and, although there appears to be a correlation between the oxygen bond-strength and catalytic activity towards the total oxidation of hydrocarbons, its relevance towards selective oxidation is seriously questioned.

C. Other Systems

(a) *Production of buta-1,3-diene.* The postulates of Mars and Van Krevelen [500] have been stated previously but, because the consequences of their model are many, they will be repeated. Basically, it was envisaged that the catalytic oxidation of hydrocarbons proceeds via the following steps:

(1) Reaction between the hydrocarbon and the oxide, to give products and a partially-reduced catalyst.
(2) Reoxidation of the reduced catalyst with gaseous oxygen to restore the catalyst to its original state.

In this model, it is obvious that the agent responsible for oxidation is an oxygen ion of the oxide lattice. This led Sachtler and De Boer [499] to speculate that the activity and selectivity of oxide catalysts towards hydrocarbon oxidation should depend on the strength of the metal-oxygen bond within the catalyst. In other words, strong bonds should produce inactive catalysts, weak bonds should yield catalysts which favour complete combustion, whilst bonds of intermediate strength should be present in catalysts for selective oxidation.

Simons *et al.* [715] attempted to confirm this theory by studying the interaction, at various temperatures, between but-1-ene, buta-1,3-diene and certain transition metal oxides. A pulse technique was employed, and the oxides chosen (MnO_2, V_2O_5, CuO, Co_3O_4, Fe_2O_3, NiO, TiO_2, SnO_2, ZnO and Cr_2O_3) are here recorded in an order such that, for the reaction

$$MO_n \rightarrow MO_{n-1} + \tfrac{1}{2}O_2 - Q_0 \text{ kcal.}$$

Q_0 increases in value from left to right.

In these experiments, Simons *et al.* [715] measured certain characteristics of the reaction over the oxides, including the maximum conversion of the feed-gas to butadiene and the so-called "T_{50}", which represented the temperature of the oxide at which 50% of the reactants (either but-1-ene or buta-1,3-diene) were converted to carbon dioxide.

It was found (see Fig. 7) that T_{50} for the oxidation of but-1-ene to CO_2 (Fig. 7(b)) and of buta-1,3-diene to CO_2 (Fig. 7(c)) indeed increased linearity with Q_0. In Fig. 7(b) and (c), the scatter appears to reflect the widely differing surface

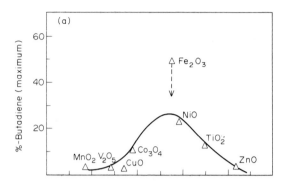

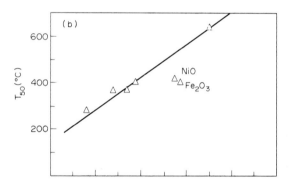

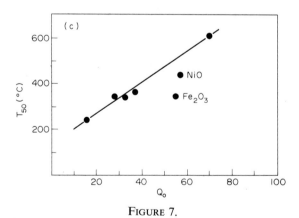

FIGURE 7.

areas of the oxides used. For example, the sample of Fe_2O_3 used had a surface area of 121·1 m² g⁻¹ compared with 2·1 m² g⁻¹ and 0·35 m² g⁻¹ for samples of titania and copper(II) oxide, respectively. These data, therefore, certainly provide good evidence for the correctness of the postulate that catalytic activity should increase with decreasing bond strength. An examination of Fig. 7(a) also seems to confirm the second theory of Sachtler and De Boer [499] that there should be a maximum in selectivity for intermediate values of Q_0. This value appears to be about 50–60 kcal. mole⁻¹ although it is probably coincidental that the Q_0 values for MoO_3 and Bi_2O_3 (components of the highly selective catalyst, bismuth molybdate) are 50 and 40 kcal. mole⁻¹, respectively.

In the course of their work, Simons et al. [715] attempted to elucidate the action of both gaseous and lattice oxygen. Their results are shown in Fig. 8(a) and (b). Fig. 8(a) shows that the conversion of but-1-ene increases linearly with its partial pressure, in the presence of bismuth molybdate, and that the fraction converted to buta-1,3-diene (90%) remains constant over a whole range of but-1-ene:O_2 compositions. The conversion of gaseous oxygen is seen to be rapid and is equivalent to the amount of but-1-ene converted if oxygen is in excess. With tin(IV) oxide, the situation is very different (Fig. 8(b)). Pure but-1-ene reacts only partially to give only 50% buta-1,3-diene and a small quantity of carbon dioxide; the rest is not accounted for. Increasing the partial pressure of oxygen increases the amount of but-1-ene converted, initially, but then the conversion decreases with decreasing but-1-ene pressure. At the lower hydrocarbon pressures, total combustion of but-1-ene was very pronounced. To Simons et al. [715], the initial low reactivity of the tin oxide surface and the sudden predominance of complete oxidation products, indicated, not the restoration of the original surface in the presence of gaseous oxygen, but the formation of a surface containing "peroxide groups".

With regard to the Mars–Van Krevelen theory [500] and the observations of Simons et al. [715], Cornaz et al. [716] had earlier observed certain phenomena which might provide a clue as to the degree to which those catalysts allow selective oxidation to proceed. Cornaz et al. [716] recorded the esr spectra of samples of anatase (TiO_2) and MoO_3, after evacuation at 500°C for 4 h, and after the subsequent addition of gaseous oxygen at room temperature. The high temperature degassing of the oxides results in partial reduction of the oxides in question and, with degassed TiO_2, two components of the resulting esr spectrum were assigned to an F-centre and to Ti^{3+} ions, both on the surface. With MoO_3, after evacuation, four-component signals were observed, one of which was assigned to an F-centre, one to surface Mo^{5+} ions, whilst the other two remained unidentified.

Addition of oxygen, in the case of anatase, caused the emission of signals containing at least four components. Cornaz et al. [716] regarded three of these signals as indicative of three types of surface coordination complexes of oxygen

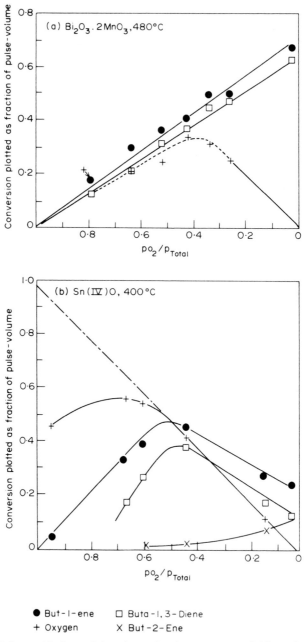

FIG. 8. Pulse oxidation of but-1-ene in the presence of bismuth molybdate (Bi : Mo = 1 : 1) and tin(IV) oxide. (Taken from Ref. [715].)

with titanium L-, P- and A-forms:

$$Ti\!=\!\!O\!\equiv\!\!O\!:\qquad Ti\!\!=\!\!\overset{\overset{..}{O}}{\underset{\underset{..}{O}}{|}}\qquad Ti\!\!-\!\!\overset{..}{O}\diagdown\overset{..}{\underset{..}{O}}$$

| L-form | P-form | A-form |

The fourth signal was believed to be somehow connected with the species O^-. With MoO_3, the addition of gaseous oxygen resulted in none of the signals observed with TiO_2. Indeed, interaction with oxygen merely served to decrease the intensity of the F-centre and Mo^{5+} signals.

Some results were also obtained by Cornaz et al. [716], with bismuth molybdate (Bi:Mo $\simeq$ 1:1) partially reduced with but-1-ene. Although the spectra obtained were similar to those from MoO_3, the work was abandoned, apparently, because of difficulty of applying esr techniques to this particular compound.

Returning to the information which can be derived from the observations of Simons et al. [715] and of Cornaz [716], it would appear that the clue to the selectivity of certain catalysts lies in the reoxidation step of the Mars–Van Krevelen mechanism. Fully oxidized tin(IV) oxide has a reasonable selectivity towards buta-1,3-diene when initially employed in the oxidation of but-1-ene and yet this selectivity disappears in the presence of gaseous oxygen. In the light of the results of Cornaz and co-workers [716], it appears that in the reoxidation step, the oxygen does not restore the catalyst to its original form, but is adsorbed on the surface in a state of enhanced reactivity. Indeed, the surface coordination complexes of oxygen observed by Cornaz et al. [716[for anatase, may also occur on reoxidation of a number of reduced oxides. Further evidence to support this idea is also forthcoming from the findings of Cornaz [716] with respect to MoO_3. In Chapter 3, it was stated that Lazukin et al. [203] had found molybdenum(VI) oxide a very selective, if not a particularly active, catalyst in the oxidation of propylene to acrolein. It is very interesting to note, therefore, that reoxidation of partially reduced MoO_3 yielded no signals of the type observed with anatase, but merely restored this oxide to its original state.

(b) Periodic-pulse studies. Niwa and Murakami [717, 718] have recently reported a technique for studying mixed-oxide catalysts which has allowed further insight into the factors which determine selectivity in hydrocarbon oxidation reactions. The technique involves the pulse-feeding of oxygen and hydrocarbon, alternately and separately, into a reactor containing the catalyst. The results thus obtained are said to approximate, more nearly, to the conditions existing during continuous-flow than to those used in pulse techniques, for example, by Simons et al. [715] and others.

Niwa and Murakami [717] initially studied the oxidation of propylene over a number of binary oxide mixtures, including unsupported Bi–Mo(1:1)–O, silica-supported Bi–Mo(1:1)–O, Bi–W(1:1)–O, Sb–Mo(2:3)–O, Sn–Sb(10:1)–O, Sn–Sb(4:1)–O, and Sn–P(10:1)–O catalysts. Pulses of oxygen + nitrogen (O-pulse) and propylene + N_2 (R-pulse) were admitted alternately to a reactor and the gaseous products were sampled and analyzed. The results of these experiments were interesting. Certain catalysts yielded acrolein on the R-pulse and carbon dioxide on the O-pulse, whilst other yielded more carbon dioxide on the R-pulse than the O-pulse.

On the basis of the product-dependence on the period, Niwa and Murakami [717] classified the catalysts studied into three groups:

    (a) Bi–Mo, Bi–W
    (b) Sn–Sb, Sb–Mo
    (c) MoO_3, Sn–P

With groups (a) and (c), combustion (to carbon dioxide) took place mainly on the O-pulse, although oxides of carbon were obtained from both types of pulse. Catalysts of group (b) yielded more combustion products on the R-pulse. It was considered that, on the R-pulse, a certain amount of hydrocarbon remained irreversibly-adsorbed on the surface of the catalyst and the pulse at which the combustion reaction proceeded was determined largely by the reactivity of the adsorbed hydrocarbon residue and the activity of the surface oxygen. In the case of catalysts such as Bi–Mo–O and Bi–W–O, it was suggested that, after the R-pulse, the surface was covered with an adsorbed hydrocarbon residue which was later attacked by gaseous O_2 (O-pulse) to yield carbon dioxide. With catalysts such as Sn–Sb–O and Sn–Mo–O, where more carbon dioxide was formed on the R-pulse, it was suggested that the surface was covered, mainly, with "active" oxygen and only a small amount of adsorbed hydrocarbon.

According to Niwa and Murakami [717], catalyst reducibility is a determining factor in the period-dependence of catalysts, combustion occurring between the irreversibly adsorbed residue and gaseous oxygen particularly with easily-reduced catalysts.

Niwa and Murakami [717] also measured the electrical conductivity of catalysts under real experimental conditions and their results suggest that acrolein formation is independent of the state of the catalyst surface since the production of this compound by the period-pulse technique was independent of the conductivity of the catalysts. Only the combustion reaction depended on the catalyst surface.

Niwa and Murakami obtained similar results for the oxidation of but-1-ene and but-2-ene to buta-1,3-diene [718].

(c) *Catalyst weight loss.* Recently, during studies of vanadium pentoxide-based catalysts Cole *et al.* [719] reported a very interesting finding which indicates an important correlation between the thermal behaviour (D.T.A., D.T.G.) of certain oxides and their catalytic properties.

Industrially, vanadium pentoxide, usually in combination with one or more components, is used extensively as a heterogeneous catalyst for reactions such as the oxidation of ortho-xylene to phthalic anhydride. For example, V_2O_5 (1–15 w %), mixed with anatase (99–85 w %), is a very versatile, highly selective and durable catalyst [720], whilst V_2O_5, promoted by only very small (<1 %) quantities of, say, zinc oxide or titanium dioxide, is claimed to be a particularly useful catalyst [721].

Since both zinc oxide and titanium oxide are believed to be excellent promoters for V_2O_5 [721] and are n-type semiconductors with comparable ionic radii ($r_{Ti}^{4+} = 0.68$Å, $r_{Zn}^{2+} = 0.74$ Å), it was very surprising to discover that, with catalyst compositions similar to those used by BASF for V_2O_5–TiO_2 catalysts [720], V_2O_5–ZnO catalysts were very unselective and not particularly active. It was in order to understand this difference more fully, that Cole et al. [719] undertook studies of vanadia-based catalysts.

Thermal studies of catalysts were initiated by the observation of colour changes when industrial V_2O_5–TiO_2 catalysts were heated or used for the partial air-oxidation of ortho-xylene. Such colour changes had been observed previously and has been ascribed by Simard et al. [722] and Vol'fson et al. [723] to the reaction

$$V_2O_5 \rightarrow V_2O_{4.34}$$

which, it was thought, would be accompanied by a weight loss.

Initially, synthetic V_2O_5–TiO_2 (anatase) mixtures were prepared, covering the composition range 100 % V_2O_5 to 100 % TiO_2, and analysed thermally on an extremely accurate thermobalance. For V_2O_5–TiO_2 mixtures, heated in air, an endothermic weight loss was noted in the temperature range 630–730°C. Despite the relatively low melting-point of V_2O_5 (690°C), this weight loss was not explicable in terms of its volatilization. The maximum weight loss occurred with binary mixtures in the composition range 20 to 25 % V_2O_5–80 to 75 % TiO_2 and was also accompanied by a gradual transformation of anatase into rutile.

The behaviour of the mixtures was further investigated by heating them in atmospheres of pure oxygen, nitrogen and hydrogen, and weight losses were again observed (Fig. 9), but the weight loss appeared to depend on the partial pressure of oxygen in the atmosphere surrounding the heated sample. In order to investigate the latter phenomenon more fully, a sample containing 30 w % V_2O_5 was heated in various mixtures of oxygen and nitrogen. At low partial pressures of oxygen (<0.4), there was a marked variation in weight-loss with partial pressure but above this concentration, the variation was minimal (Fig. 10).

Mass spectrometric analysis revealed that, during weight-loss, oxygen was

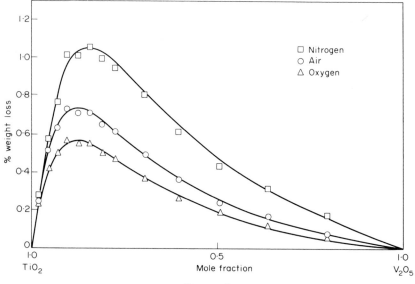

FIGURE 9.

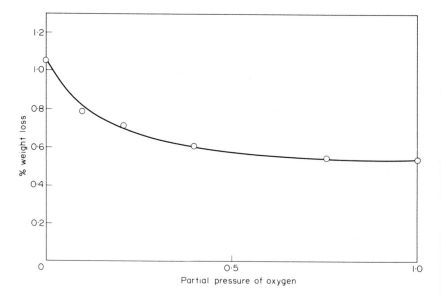

FIGURE 10.

evolved from the catalysts, the amount of which depended on the fraction of V_2O_5 in the mixtures (Fig. 11). One obvious reaction involving the evolution of oxygen from V_2O_5 is

$$V_2O_5 \rightarrow V_2O_{5-n} + \frac{n}{2}O_2$$

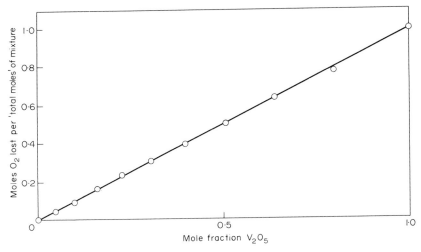

FIGURE 11.

which is accompanied by a change in the oxidation state of vanadium. V_2O_5–TiO_2 (10, 30, 50, 80% (w/w) V_2O_5) samples were heated, therefore, in nitrogen and then analysed for their reduced-vanadium content. These results are shown in Fig. 12. The curve is similar in shape to the weight-loss curve and agrees well with the weight loss predicted on the basis of the above equation and the mixture composition.

Although the observations have been made under conditions apparently remote from those encountered in catalytic oxidation reactions involving V_2O_5, evidence has been obtained which suggests a connection between the thermal behaviour of certain catalysts and their activity with respect to certain reactions. Thus the curves of % weight loss against mixture composition, show a maximum in the region 0·1 to 0·2 mole fraction V_2O_5, a composition range preferred by BASF [720]. Vanhove and Blanchard [724] have also obtained data regarding the selectivity (towards phthalic anhydride) of various V_2O_5–TiO_2 mixtures used in the oxidation of o-xylene and, although these exhibit two maxima, one maximum for low V_2O_5 mixtures coincides with the maximum weight loss in the present work (Fig. 13). Finally, investigations of V_2O_5–ZnO mixtures initially have shown that pure zinc oxide loses weight in the temperature range 200–450°C and similar behaviour has been found in the presence of

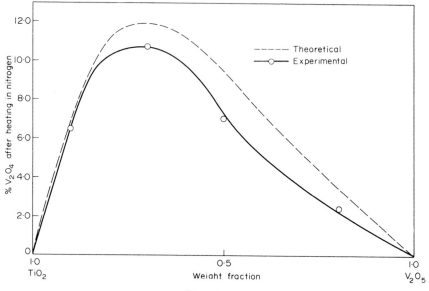

FIGURE 12.

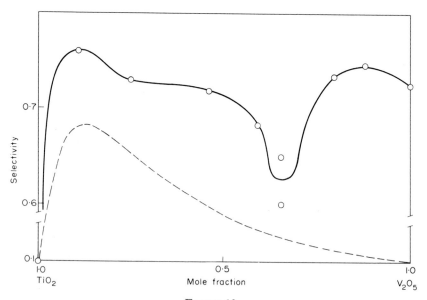

FIGURE 13.

V_2O_5. The heating atmosphere has also been shown to have no effect on the weight loss, which is directly proportional to the amount of zinc oxide in the sample (Fig. 14). In this case, oxygen from the V_2O_5 lattice is not participating in the weight loss and it is interesting to note that V_2O_5–ZnO mixtures are generally unsatisfactory in partial oxidation experiments.

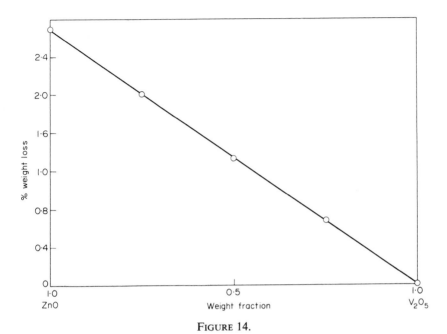

FIGURE 14.

In conclusion, although the relation between oxygen weight loss and the preferred catalyst composition for selective oxidation may be fortuitous for the systems examined, the evidence so far obtained suggests a symbiotic interaction between V_2O_5 and certain other transition metal oxides, the effect on V_2O_5 being to labilize the surface oxide ions *even in the presence of gaseous oxygen*.

6.3. Conclusions

Extensive studies of the selective oxidation of hydrocarbons (other than ethylene) have revealed surprisingly few facts of general applicability. Experimental evidence indicates that all selective oxidation catalysts are composed of at least two oxide components, one responsible for activity and the other for selectivity but it is still not known whether, for example, both cationic centres participate in surface processes or whether the role of the activating component consists only in modifying the oxygen–metal bond in the base catalyst [719].

In the case of so-called allylic oxidations (the oxidation and ammoxidation of propylene and isobutene to yield acrolein, acrylonitrile, etc., the oxidation of the butenes) certain facts have been established which may be accepted as universal:

(1) The first step in the reaction involves the dissociative chemisorption of the olefin with the abstraction of an α-hydrogen, yielding a symmetrical allylic intermediate.
(2) This first step is rate-determining.
(3) In many cases, the rate of formation of reaction products is equal to the rate of reduction of the catalyst in the absence of gaseous oxygen, thus favouring the theory of Mars and Van Krevelen [500].

Only in the case of one catalyst, bismuth molybdate, has a complete picture of oxidation catalysis emerged, and this has been achieved only after almost seven years' work by numerous investigators.

REFERENCES

1. Margolis, L. Ya., *Adv. Catalysis*, **14**, 429 (1963).
2. Sampson, R. J., and Shooter, D., *Oxid. Combust. Rev.*, **1**, 225 (1965).
3. Voge, H. H., and Adams, C. R., *Adv. Catalysis*, **17**, 151 (1967).
4. Sachtler, W. M., *Catal. Rev.*, **4**, 27 (1970).
5. Beranek, L., *Colln. Czech. chem. Commun.*, **32**, 489 (1967).
6. Satterfield, C. N., and Sherwood, T. K., "The Role of Diffusion in Catalysis", Adison-Wesley Publishing Co. Inc., Reading, Mass., 1963.
7. Satterfield, C. N., "Mass Transfer in Heterogeneous Catalysis", M.I.T. Press, Cambridge, Mass., and London, 1970.
8. Wheeler, A., *Adv. Catalysis*, **3**, 249 (1951).
9. Butt, J. B., *Chem. Engng Sci.*, **21**, 275 (1966).
10. Van der Vusse, J. G., *Chem. Engng Sci.*, **21**, 645 (1966).
11. Mingle, J. O., and Smith, J. M., *A.I.Ch.E.Jl*, **7**, 243 (1961).
12. Carberry, J. J., *Chem. Engng Sci.*, **17**, 675 (1962).
13. Beek, J., *A.I.Ch.E.Jl.*, **7**, 377 (1961).
14. Hlavacek, V., and Marek, M., *In* "Chemical Reaction Engineering, Proceedings of the Fourth European Symposium (1968)" (R. Jottrand, ed.), p. 207, Pergamon, Oxford, 1971.
15. Roberts, C. W., *Chem. Engng Sci.*, **27**, 1409 (1972).
16. Tartarelli, R., *Chimica Ind.*, *Milano*, **49**, 620 (1967).
17. Tartarelli, R., *Chimica Ind.*, *Milano*, **50**, 556 (1968).
18. Tartarelli, R., and Capovani, M., *Chimica Ind.*, *Milano*, **50**, 1318 (1968).
19. Tartarelli, R., and Morelli, F., *J. Catal.*, **11**, 159 (1968).
20. Tartarelli, R., Cioni, S., and Capovani, M., *J. Catal.*, **18**, 212 (1970).
21. Ferraiolo, G., and Beruto, D., *Quad. Ing. Chim. Ital.*, **5**, 165 (1969).
22. Wicke, E., *Mem. Soc. Roy. Sci.*, *Liege*, **1**, 211 (1971).

23. Boreskov, G. K. *In* "Porous Structure Catalytic Transport Processes in Heterogeneous Catalysis, Symposium 1968" (G. K. Boreskov, ed.), p. 1, Akademiai Kaido, Budapest (1972).
24. Shooter, D. *In* "Comprehensive Chemical Kinetics", Vol. I, p. 222 (C. H. Bamford, and C. F. H. Tipper, eds), Elsevier Publishing Co., Amsterdam, 1969.
25. Levenspiel, O., "Chemical Reaction Engineering; An Introduction to the Design of Chemical Reactors", John Wiley, New York, 1962.
26. Kotera, Y., and Miki, Y., *Kogyo Kagaku Zasshi*, **73**, 259 (1970).
27. Aliev, V. S., Efendiev, R. M., and Lieberman, E. S., *Azerb. Khim. Zh.* (4), 29 (1968); *Chem. Abstr.*, **70**, 67518p.
28. Urwin, D. *In* "Proceedings of the International Symposium on Surface Area Determination, Bristol, 1969" (D. Everett and R. H. Ottewill, eds), p. 397, Butterworths Scientific Publications, London, 1970.
29. Karnaukhov, A. P., and Buyanova, N. E. *In* "Proceedings of the International Symposium on Surface Area Determination, Bristol, 1969" (D. Everett and R. H. Ottewill, eds), p. 165, Butterworths Scientific Publications, London, 1970.
30. Farey, M. C., and Tucker, B. G., *Anal. Chem.*, **43**, 1307 (1971).
31. Buyanova, N. E., Karnaukhov, A. P., and Chernyavskaya, O. N., *Gazov. Khromatogr.* (9), 113 (1969).
32. Buyanova, N. E., Kuznetsova, E. V., and Borisova, M. S., *Zav. Lab.*, **35**, 154 (1969).
33. Bliznakov, G. M., Bakardjiev, I. V., and Gocheva, E. M., *J. Catal.*, **18**, 260 (1970).
34. Voge, H. H., ACS No. 76 Vol. II, "Oxidation of Organic Compounds", p. 242, Am. Chem. Soc., Washington, 1968.
35. Kolchin, I. K., Bobkov, S. S., and Margolis, L. Ya. *Neftekh.*, **4**, 301 (1964).
36. Little, L. H., "I.r. Spectra of Absorbed Species", Academic Press, New York, 1966.
37. Hair, M. L., "I.r. Spectroscopy in Surface Chemistry", M. Dekker, New York, 1967.
38. Amberg, C. H. *In* "The Gas-solid Interface" (E. A. Flood, ed.), M. Dekker, New York, 1967.
39. Basila, M. R., *Appl. Spectros. Revs*, (E. G. Braine Jr., ed.), **1**, p. 289, M. Dekker, New York, 1967.
40. Dent., A. L., and Kokes, R. J., *J. Am. chem. Soc.*, **92**, 1092, 6709 (1970).
41. Naito, S., Kondo, T., Ichikawa, M., and Tamaru, K., *J. phys. Chem.*, **76**, 2184 (1972).
42. Whitney, A. G., and Gay, I. D., *J. Catal.*, **25**, 176 (1972).
43. Ishida, S., and Doi, Y., *Chubu Kogyo Daigaku Kyo*, **5**, 161 (1969).
44. Shvets, V. A., and Kazanskii, V. B., *J. Catal.*, **25**, 123 (1972).
45. Krylov, O. V., Pariiskii, G. B., and Spiridonov, K. N., *J. Catal.*, **23**, 301 (1971).
46. Van Reijen, L. L., *Ber. Bunsenges. Phys. Chem.*, **75**, 1046 (1971).
47. McCain, C. C., Gough, G., and Godin, G. W., *Nature, Lond.*, **198**, 989 (1963).
48. Okamoto, Y., Happel, J., and Koyama, H., *Bull. chem. Soc. Japan*, **40**, 2333 (1967).
49. Delglass, W. N., Hughes, T. R., and Fadley, C. S., *Catal. Rev.*, **4**, 179 (1970).
50. Penelle, R., *Métaux*, **43**, 339 (1968).
51. Terenin, A. N., *Metody Issled. Katal. Katal. Reakts.*, **4**, 56 (1971); *Chem. Abstr.*, **77**, 10063b.

52. Landau, R., Brown, D., Russell, J. L., and Kollar, J., *Proc. 7th World Petrol. Cong.*, **5**, 67 (1967).
53. Lefort, T. E., U.S. Pat., 1,998,878 (1935).
54. Asinger, F., "Mono-olefins Chemistry and Technology" (translated by B. J. Hazzard), p. 568, Pergamon Press, London, 1968.
55. Bodson, J. J., *Ind. Chim. Belge*, **32**, 880 (1967).
56. Halcon International Inc., Neth. Appl., 6,412,836 (1965).
57. Echigoya, E., and Osberg, G. L., *Can. J. chem. Engng*, **38**, 43 (1960).
58. Boreskov, G. K., Vasilevich, L. A., Gur'yanova, R. N., Kernerman, C. Sh., Slin'ko, M. G., Filippova, A. G., and Chesnokov, B. B., *Kinet. Katal.*, **3**, 214 (1962).
59. Wasilewski, E., and Kubica, Z. *In* "Conf. Chem. Chem. Process. Natur. Gas, Plenary Lect., Budapest 1965" (M. Freund, ed.), p. 447, Akademiai Kaido, Budapest, 1968; *Chem. Abstr.*, **70**, 47183X.
60. Erdoelchemie G.m.b.H., Neth. Appl., 6,605,594 (1966).
61. Keith, C. D., Hinden, S. G., and Galen, L. A., U.S. Pat., 3,420,784 (1969).
62. Hirasa, K., Noguchi, I., and Hirayama, T., *Kogyo Kagaku Zasshi*, **69**, 36, 41, 45, 52, 55 (1966); *Chem. Abstr.*, **65**, 7121f, g, h.
63. Degeilh, R. *In* "Osn. Predvideniya Katal. Deistviya, Tr. Mezhdunar, Kong. Katal. 4th 1968" (Ya. T. Eidus, ed.) (Publ. 1970), **2**, 69 Nauka, Moscow.
64. Mistrik, E. J., Ondrus, I., and Gregor, F., Czech. Pat., 120251 (1966).
65. Khasin, E. I., Kagan, N. M., Chizhik, S. P., and Mezis, L. M., *Porosh. Met.*, **11**, 78 (1971); *Chem. Abstr.* **76**, 18273v.
66. Vasilevich, L. A., Boreskov, G. K., Gur'yanova, R. N., Ryzhak, I. A., Filippova, A. G., and Frolkina, I. T., *Kinet. Katal.*, **7**, 525 (1965).
67. Presland, A. E. B., and Trimm, D. L., *Micron*, **1**, 52 (1969).
68. Harriott, P., *J. Catal.*, **21**, 56 (1971).
69. Engelhard Minerals and Chem. Corp., Br. Pat., 1,129,190 (1968).
70. Engelhard Industries Inc., Br. Pat., 1,103,678 (1968).
71. Calcagno, B., Ferlazzo, N., and Ghirgha, M., Ger. Offen., 1,958,586 and 1,958,597 (1970).
72. B.A.S.F. A.-G. ,Neth. Appl., 6,503,000 (1965).
73. Brown, D., Porcelli, J. V., and Flaster, E. R., Ger. Offen., 2,034,358 (1971).
74. Allied Chem. Corp., Fr. Pat., 1,413,213 (1965).
75. Halcon Internat. Inc., Neth. Appl., 6,606,498 (1966).
76. Farbenfabriken Bayer A.-G./Erdoelchemie G.m.b.H., Neth. Appl., 6,507,088 (1966).
77. I.C.I. Ltd., Neth. Appl., 6,606,581 (1966).
78. Brown, D., and Saffer, A., Ger. Offen., 1,260,451 (1968).
79. Janda, J., and Koma, J., Czech. Pat., 122,482 (1967).
80. Fujii, T., Matsuda, K., Kumazawa, J., and Kiguchi, I., Jap. 71, 19,606 (1971).
81. Nielsen, R. P., Ger. Offen., 2,159,346 (1972).
82. Shin, Y. J., and Shirozaki, T., Jap., 71, 40,921 (1971).
83. Arai, Y., Jap., 71, 19,605 (1971).
84. Stiles, A. B., U.S. Pat., 3,560,530 (1971).
85. Cusumano, J. A., Ger. Offen., 2,062,190 (1971).
86. Kucirka, J. F., U.S. Pat., 3,654,318 (1972).
87. Piccinini, C., Morelli, M., and Rebora, P., S. African Pat., 70 05,849 (1971).
88. Flaster, E. R., and Porcelli, J. V., Ger. Offen., 2,219,748 (1972).
89. Halcon International Inc., Brit. Amended, 1,020,759 (1969).

90. Chan, P. E., Ger. Offen., 2,141,574 (1972).
91. Wasilewski, J., *Zesz. Nauk, Inst. Ciezkiej Syntezy Organiczrej Blackowni Slask.*, **2**, 128 (1970); *Chem. Abstr.*, **74**, 31674e.
92. Janda, J., and Kluckovsky, P., *Chem. Zvesti*, **20**, 267 (1966); *Chem. Abstr.*, **65**, 3817a.
93. Kaliberdo, L. M., Dorogova, V. B., and Pashegorova, V. S., *Neftekh.*, **11**, 719 (1971).
94. Ostrovskii, V. E., Kul'kova, N. V., Lopatin, V. L., and Temkin, M. I., *Kinet. Katal.* (Engl. Transl.), **3**, 160 (1962).
95. Ostrovskii, V. E., and Temkin, M. I., *Kinet. Katal.*, **7**, 529 (1966).
96. Rudnitskii, L. A., Kul'kova, N. V., and Temkin, M. I., *Metody Issled. Katal. Katal. Reakts. Akad. Nauk SSSR, Sib. Otd., Inst. Katal.*, **3**, 145 (1965); *Chem. Abstr.* **67**, 57560v.
97. Rudnitskii, L. A., and Kul'kova, N. V., *Kinet. Katal.*, **7**, 759 (1966).
98. Spath, H., and Torkar, K., *J. Catal.*, **26**, 163 (1972).
99. Spath, H., Tomazin, G. S., Wurm, H., and Torkar, K., *J. Catal.*, **26**, 18 (1972).
100. Kim, S. Y., Kim, Y. R., Li, G. S., and Kim, G. U., *Choson Minjujuui Inmin Konghwaguk Kwahagwan Tongbo* (5), 3 (1972); *Chem. Abstr.*, **76**, 117846r.
101. Carberry, J. J., Kuczynski, G. C., and Martinez, E., *J. Catal.*, **26**, 247 (1972).
102. Carberry, J. J., Kuczynski, G. C., and Martinez, E., Ger. Offen., 2,205,812 (1972).
103. Kuczynski, G. C., Carberry, J. J., and Martinez, E., *J. Catal.*, **28**, 39 (1973).
104. MacDonald, R. W., and Hayes, J. E., *J. Catal.*, **15**, 301 (1969).
105. Kenson, K. E., and Lapkin, M., *J. phys. Chem.*, **74**, 1493 (1970).
106. Wichterlova, B., *Chem. Listy*, **66**, 1178 (1972); *Chem. Abstr.*, **78**, 28900c.
107. Twigg, G. H., *Proc. R. Soc.*, *A***188**, 92 (1946); *Trans. Faraday Soc.*, **42**, 284 (1946).
108. Margolis, L. Ya., and Roginskii, S. Z., *Izv. Akad. Nauk SSSR, Otdel. Khim. Nauk* p. 281 (1956); *Chem. Abstr.*, **51**, 199h.
109. Alfani, F., and Carberry, J. J., *Chimica Ind., Milano*, **52**, 1192 (1970).
110. Rastaturin, V. A. *Zh. Priklad. Khim.*, **43**, 1343 (1970); *Chem. Abstr.* **73**: 91958.
111. Hartwig, I., and Bathory, J., *Magy. Asvanyolaj Foldgaz kiserl. Intez. Kozlem.*, **8**, 75, 98 (1967); *Chem. Abstr.*, **68**, 95089b.
112. Hartwig, I., and Bathory, J., *Z. phys. Chem.*, **55**, 208 (1967); *Chem. Abstr.*, **67**, 99399e.
113. Ide, Y., Takagi, T., and Keii, T., *Nippon Kagaku Zasshi*, **86**, 1249 (1965); *Chem. Abstr.*, **65**, 12149g.
114. Neufeld, M. L., and Blades, A. T., *Can. J. Chem.*, **41**, 2956 (1963).
115. Benson, S. W., *J. chem. Phys.*, **40**, 105 (1964).
116. Rinker, R. G., Shah, G. B., and Chock, E. P., *Ind. Eng. Chem. Fundam.*, **10**, 13 (1971).
117. Ayame, A., Kano, H., and Kanazuka, T., *Muroran Kogyo Daigaku Kenkyu Hokou*, **7**, 55 (1970); *Chem. Abstr.*, **74**, 125288j.
118. Balogh, A., and Bathory, J., *Acta. Chim.* (*Budapest*), **58**, 287 (1967).
119. Czanderna, A. W., *J. Colloid Interface Sci.*, **22**, 482 (1966).
120. Gerei, S. V., Kholyavenko, K. M., and Rubanik, M. Ya., *Probl. Kinet. Katal., Akad. Nauk SSSR*, **12**, 118 (1968).
121. Kagawa, S., Tokunaga, H., and Seiyama, T., *Kogyo Kagaku Zasshi*, **71**, 775 (1968); *Chem. Abstr.*, **69**, 95565z.

122. Feller-Kniepmeier, M., Feller, H. G., and Titzenhaler, E., *Ber. Bunsenges. Phys. Chem.*, **71**, 606 (1967).
123. Degeilh, R. *In* "Proceedings of the Fourth Internal Congress on Catalysis, 1968" (D. A. Kazanaskii, ed.), **2**, p. 35, Akademiai Kaido, Budapest, 1971.
124. Bradshaw, A. M., Engelhardt, A., and Menzel, A., *Ber. Bunsenges. Phys. Chem.*, **76**, 500 (1972).
125. Otto, E. M., *J. electrochem. Soc.*, **113**, 643 (1966).
126. May, M. W., and Linnett, J. W., *J. Catal.*, **7**, 324 (1967).
127. Kagawa, S., Kono, K., Futata, H., and Seiyama, T., *Kogyo Kagaku Zasshi*, **74**, 819 (1971); *Chem. Abstr.*, **75**, 48150z.
128. Liberti, G., Mattera, A., Pedretti, F., Pernicone, N., and Saottini, S., *Atti, Accad. naz. Lincei, Cl. Sci. Fis., Mat. Natur., Rend.*, **52**, 392 (1972).
129. Czanderna, A. W., *J. phys. Chem.*, **68**, 1765 (1964).
130. Kummer, J. T., *J. phys. Chem.*, **63**, 460 (1959).
131. Sandler, T. L., and Durigon, D. D., *J. phys. Chem.*, **69**, 4201 (1965).
132. Sandler, T. L., and Durigon, D. D., *J. phys. Chem.*, **70**, 3881 (1966).
133. Ohashi, H., Kano, H., Sato, N., and Okamoto, G., *Kogyo Kagaku Zasshi*, **69**, 997 (1966); *Chem. Abstr.*, **69**, 110241m.
134. Sato, N., and Seo, M., *Nature, Lond.*, **216**, 361 (1967).
135. Sato, N., and Masahiro, S., *J. Catal.*, **24**, 224 (1972).
136a. Sato, N., and Seo, M., *Denki Kagaku*, **38**, 649 (1970).
136b. Seo, M., and Sato, N., *Denki Kagaku*, **38**, 768 (1970); *Chem. Abstr.*, **75**, 10546b.
137. Lilov, I. V., Bliznakov, G., and Syrnev, L., *Kinet. Katal.*, **10**, 930 (1969).
138. Schultze, G. R., and Thiele, H., *Erdoel Kohle*, **5**, 552 (1952).
139. Charmon, H. B., Dell, R. M., and Teale, S. S., *Trans. Faraday Soc.* **59**, 453 (1963).
140. Herzog, W., *Ber. Bunsenges. Phys. Chem.*, **74**, 216 (1970).
141. Sandler, Y. L., and Hickham, W. M. *In* "Proceedings of the Third International Congress on Catalysis, Amsterdam, 1964" (W. M. H. Sachtler, G. C. A. Schuit and P. Zwietering, eds), North-Holland Publishing Co., Amsterdam, 1965.
142. Woodward, J. W., Lindgren, F. G., and Corcoran, W. H., *J. Catal.*, **25**, 292 (1972).
143. Kemball, C., and Patterson, W. R., *Proc. R. Soc.*, *A***270**, 219 (1962).
144. Boutry, P., and Montarnal, R., Fr. Pat., 1,527,716 (1968).
145. Institut Francais du Petrole, des Carburants et Lubrifiants, Fr. Pat. 1,536,185 (1968).
146. Boutry, P., and Montarnal, R., Fr. Pat., 1,568, 742 (1969).
147. Capp., C. W., and Harris, B. W., Br. Pat., 1,142,897 (1969).
148. Elek, L. F., Risch, A. P., Evnin, A. B., Rabo, J. A., and Kavarnos, S. J., Ger. Offen., 2,042,815 (1971).
149. Gerberich, H. R., and Hall, W. K., U.S. Pat., 3,534,093 (1970).
150. Wattimena, F., U.S. Pat., 3,293,291 (1966).
151. McClain, D. M., Heller, C. A., and Mador, I. L., Ger. Offen., 2,162,779 (1972).
152. Nakanishi, Y., Kurata, N., and Okuda, Y., Jap., 71 06,763 (1971).
153. Boudart, P., and Montarnal, R., Fr. Addn., 2,097,219 (to Fr., 2,058,817) (1972).
154. Patterson, W. R., and Kemball, C., *J. Catal.*, **2**, 465 (1963).
155. Gerberich, H. R., and Hall, W. K., *Nature, Lond.*, **213**, 1120 (1967).
156. Gerberich, H. R., Cant, N. W., and Hall, W. K., *J. Catal.*, **16**, 204 (1970).
157. Cant, N. W., and Hall, W. K., *J. Catal.*, **16**, 220 (1970).
158. Moss, R. L., and Thomas, D. H., *J. Catal.*, **8**, 151, 162 (1967).

Chem., No. 21, 109 (1971); *Chem. Abstr.*, **75**, 53735y; (b) Paryjczak, T., Lewicki, A., and Falak, B., *Zesz. Nauk. Politech. Lodz. Chem.*, No. 21, 119 (1971); *Chem. Abstr.*, **75**, 53554p.

228. Tan, S., Moro-oka, Y., and Ozaki, A., *J. Catal.*, **17**, 125, 132 (1970).
229. Buiten, J., *J. Catal.*, **10**, 188 (1968).
230. Lazukin, V. I., Rubanik, M. Ya., Zhigailo, Ya. V., Kurganov, A. A., and Buteiko, Zh. F., *Katal. Katal., Akad. Nauk SSR, Resp. Mezhvedom. Sb.* No. 2, 50 (1966); *Chem. Abstr.*, **66**, 75620y.
231. Pluta, J., and Blasiak, E., *Przem. Chem.*, **47**, 203, 273 (1968); *Chem. Abstr.*, **69**, 18473r and **70**, 28336h.
232. Gerei, S. V., Kholyavenko, K. M., Baryshevskaya, I. M., Chernukhina, N. A., and Lazukin, V. I., *Metody Issled. Katal. Katal. Reakst. Sb.* **2**, 334 (1965); *Chem. Abstr.*, **65**, 17905d.
233. Firsova, A. A., Khovanskaya, N. N., Isyganov, A. D., Suzdalev, I. P., and Margolis, L. Ya., *Kinet. Katal.* (Eng. transl.), **12**, 708 (1971).
234. Orlov, A. N., Lygin, V. I., and Kolchin, I. K., *Kinet. Katal.* (Eng. transl.), **13**, 726 (1972).
235. Barclay, J. L., Bethell, J. R., Bream, J. B., Hadley, D. J., Jenkins, R. H., Stewart, D. G., and Wood, B., *Br. Pat.*, 864,666 (1960); Bream, J. B., Hadley, D. J., Barclay, J. L., and Stewart, D. G., *Br. Pat.*, 876,446 (1961); Barclay, J. L., and Hadley, D. J., *Br. Pat.*, 902,952 (1962); Bethell, J. R., and Hadley, D. J., U.S. Pat., 3,094,565 (1963).
236. Belousov, V. M., and Gershingorina, A. V., *Kinet. Katal.* (Eng. transl.), **12**, 541 (1971).
237. Lazukin, V. I., Rubanik, M. Ya., Zhigailo, Ya. V., and Kurganov, A. A., *Katal. Katal. Akad. Nauk Ukr. SSR, Resp. Mezhvedom. Sb.*, No. 3, 54 (1967); *Chem. Abstr.*, **68**, 86799f.
238. Godin, G. W., McCain, C. C., and Porter, E. A. *In* "Proceedings of the Fourth International Congress on Catalysis, Moscow, 1968" (D. A. Kazanaskii, ed.), Akademiai Kaido, Budapest, 1971.
239. Belousov, V. M., and Gershingorina, A. V., *Kinet. Katal.* (Eng. transl.), **11**, 942 (1970).
240. Krylova, A. V., Derlyukova, L. E., and Margolis, L. Ya., *Dokl. Akad. Nauk SSSR* (Eng. Transl.), **178**, 79 (1968); *Probl. Kinet. Katal.*, **14**, 180 (1970).
241. Suzdalev, I. P., Firsova, A. A., Aleksandrov, A. U., Margolis, L. Ya., and Baltrunas, D., *Dokl. Akad. Nauk SSSR* (Eng. Transl.), **204**, 408 (1972).
242. Wakabayashi, K., Kamiya, Y., and Ohta, N., *Bull. Chem. Soc. Japan*, **40**, 2172 (1967).
243. Wakabayashi, K., Kamiya, U., and Ohta, N., *Bull. Chem. Soc. Japan*, **41**, 2776 (1968).
244. Roginskaya, Yu. E., Dublin, D. A., Stroeva, S. S., Kul'kova, N. V., and Gel'bshtein, A. I., *Kinet. Katal.*, **9**, 1143 (1968).
245. Dewing, J., Barrett, C., and Rooney, J. J., *Ger. Offen.*, 1,903,617 (1969).
246. Callahan, J. L., and Gertisser, B., U.S. Pat., 3,198, 750 (1965) and U.S. Pat., 3,308, 151 (1967).
247. Simons, Th. G. J., Houtman, P. N., and Schuit, G. C. A., *J. Catal.*, **23**, 1 (1971).
248. Morita, Y., Soga, S., and Furuyama, T., *Waseda Daigaku Rikogaku Kenkyusho Hokoku* (53), 118 (1971); *Chem. Abstr.*, **76**, 104, 263y.
249. Grasselli, R. K., and Callahan, J. L., *J. Catal.*, **14**, 93 (1969).
250. Grasselli, R. K., Suresh, D. D., and Knox, K., *J. Catal.*, **18**, 356 (1970).

251. Grasselli, R. K., and Suresh, D. D., *J. Catal.*, **25**, 273 (1972).
252. Lazukin, V. I., Rubanik, M. Ya., Zhighailo, Ya. V., and Kurganov, A. A., *Katal., Katal.*, **4**, 70 (1968).
253. Ohdan, K., Umemura, S., and Yamada, K., *Kogyo Kagaku Zasshi*, **72**, 2373, 2495, (1969).
254. Ohdan, K., Umemura, S., and Yamada, K., *Kogyo Kagaku Zasshi*, **72**, 2376 (1969).
255. Ohdan, K., *J. Sci. Hiroshima U.*, Ser. A-2, **34**, (3), 317 (1970); *Chem. Abstr.*, **76**, 104261w.
256. Yamada, K., Umemura, S., Ohdan, K., Miki, K., Arima, Y., Kidaka, M., Bando, Y., Fukuda, K., and Sawazi, M., Ger. Offen., 2,136,765 (1972).
257. Badische Anilin und Sodafrabrik A.-G., Fr. Dem., 2,000,138 (1969).
258. Callahan, J. L., and Gertisser, B., Ger. Pat., 1,265,713 (1968).
259. Deutsche Erdoel A.-G., Fr. Pat., 1,558,233 (1969).
260. Shell Internat. Res. Maatsch. N.V., Neth. Appl., 6,605,325 (1966).
261. Engelbach, H., Krabetz, R., and Buechler, G., Ger. Offen., 1,920,247 (1970).
262. Societe Nationale des Petroles d'Aquitaine, Neth. Appl., 6,511,989 (1966).
263. (a) Furuya, A., Jap., 69 25,049 (1969); (b) Jap., 69 25,050 (1969); (c) Jap., 69 25,051 (1969).
264. Ikeda, T., Ishii, H., and Nakano, T., Jap., 69 25,046 (1969).
265. Ito, H., Nakamura, S., and Inoue, H., Fr. Pat., 1,579,839 (1969).
266. Ito, H., Nakamura, S., and Nakano, T., Ger. Offen., 2,000,426 (1970).
267. (a) Ito, H., Inoue, H., and Nakamura, Y., Jap., 69 25,047 (1969); (b) Jap., 69 25,046 (1969).
268. Koch, T. A., U.S. Pat., 3,387,038 (1968).
269. Mitsubishi Rayon Co. Ltd., Fr. Pat., 1,447, 982 (1966).
270. Nakayama, Y., Ogawa, Y., and Asao, S., Jap., 70 08,207 (1970).
271. Ono, I., and Iikuni, T., Jap., 68 13,609 (1968).
272. Nakayama, Y., Ogawa, Y., and Uchida, T., Jap., 70 07,290 (1970).
273. Nemec, J. W., and Schlaefer, F. W., Fr. Pat., 1,494,431 (1967).
274. Nippon Kagaku Co. Ltd., Fr. Pat., 1,514,167 (1968).
275. Parks-Smith, D. G., Br. Pat., 1,146,870 (1969).
276. Rohm und Haas G.m.b.H., Br. Pat., 1,038,274 (1966).
277. Rohm und Haas G.m.b.H., Fr. Pat., 1,554,240 (1969).
278. (a) Takenaka, S., and Yamaguchi, G., Jap., 69 06,246 (1969); (b) Jap., 69 06, 245 (1969).
279. Shell Internat. Res. Maatsch. N.V., Neth. Appl., 6,511,988 (1966).
280. Societe Nationale des Petroles d'Aquitaine, Neth. Appl., 6,511, 736 (1966).
281. Uda, K., Sakurai, A., and Abe, I., Jap., 69 04,771 (1969).
282. Sudo, M., Kita, T., and Tamayori, K., Jap., 68 10,606 (1968).
283. Takayama, Y., and Ikeda, T., Jap. 69 08,990 (1969).
284. Tokutaniyama, A., Kato, T., and Baison, K., Jap., 69 13,130 (1969).
285. Takayama, Y., Nakayama, Y., and Yoshizawa, K., Jap. 68 08,806 (1968).
286. (a) Hirota, K., Kashiwabara, H., and Nakamura, T., Jap. 68 23,925 (1968); (b) Jap. 68 23,925 (1968); (c) Jap. 68 23,927 (1968).
287. Takenaka, S., Kido, Y., Shimabara, T., and Ogawa, M., Ger. Offen., 2,038,749 (1971).
288. Yamagishi, K., Sakakibara, K., and Yasube, I., Jap., 70 06,142 (1970).
289. Takenaka, S., and Yamaguchi, G., Jap., 69 05,855 (1969).
290. Ohara, T., Ueshima, M., Nishinomiya, H., and Yanigisawa, I., Ger. Offen., 2,125,032 (1971).

291. Shiraishi, T., Kishiwada, S., Shimizu, S., Honmaru, S., Ichihashi, H., and Nagaska, Y., Ger. Offen., 2,133,110 (1972).
292. Callahan, J. L., and Gertisser, B., Ger. Pat., 1,265,731 (1968).
293. Koberstein, E., Luessling, T., Noll, E., Suchsland, H., and Weigert, W., Ger. Offen., 2,049,583 (1972).
294. Levy, L. B., Ger. Offen., 2,116, 171 (1971).
295. Ube Industries Ltd., Fr. Pat., 2,075,450 (1971).
296. Yamada, K., Umemura, S., Nagai, S., Ohdan, K., and Nakamura, I., Ger. Offen., 2,103,272 (1972).
297. Yamada, K., Umemura, S., Ohdan, K., Hidaka, M., and Fukuda, K., Ger. Offen., 2,062, 025 (1971).
298. Lane, R. E., U.S. Pat., 3,649,562 (1972).
299. Ono, I., Iiguni, T., and Akashi, M., Jap., 72 38,411 (1972).
300. Watanabe, Y., Sugihara, T., Takagi, K., Imanari, M., and Nojiri, N., Ger. Offen., 2,165,335 (1972).
301. Yashino, T., Saito, S., Sofugawa, Y., and Sasaki, T., Jap., 72 19,764 (1972).
302. Kita, T., Yamada, K., and Ishii, C., Jap., 72 40,774 (1972).
303. Grasselli, R. K., Miller, A. F., and Hardman, A. F., Ger. Offen., 2,203, 710 (1972).
304. Levy, L. B., Ger. Offen., 2,004,049 (1971).
305. Zhiznevskii, V. M., Tolopko, D. K., and Fedevich, E. V., Neftekh., 8, 68 (1968); Chem. Abstr., 69, 66646w.
306. Wakabayashi, K., and Kamiya, Y., Bull. Chem. Soc. Japan, 40, 401 (1967).
307. Golodets, G. I., Pyatnitskii, Yu. I., and Il'chenko, N. I., Dokl. Akad. Nauk SSSR (Engl. transl.), 196, 70 (1971).
308. Ohdan, K., Umemura, S., and Yamada, K., Kogyo Kagaku Zasshi, 73, 441 (1970).
309. Kuchmii, S. Ya., Gerei, S. V., and Ghorokhovatskii, Ya. B., Dopov. Akad. Nauk Ukr. SSR, Ser. B, 33, 1100 (1970); Chem. Abstr., 76, 85316z.
310. Ishikawa, T., Nishimura, T., Hayakawa, T., and Takehira, K., Bull. Japan Petrol. Inst., 13, 67 (1971).
311. Idol, J. D. Jr., Callahan, J. L., and Foreman, R. W., U.S. Pat., 2,881,213 (1959).
312. Callahan, J. L., Milberger, E. C., and Utter, R. E., U.S. Pat., 3,172, 909 (1965).
313. Komuro, I., and Koshikawa, T., Jap., 72 00,044 (1972).
314. Honda, M., Takana, K., and Watanabe, I., Ger. Offen., 2,112,938 (1972).
315. W. R. Grace and Co., Br. Pat., 1,112,638 (1968).
316. Krabetz, R., Willersinn, C. H., Engelbach, H., Wistuba, H., Hebert, U., and Frey, W., Ger. Offen., 2,056,614 (1972).
317. Bethell, J. R., and Hadley, D. J., U.S. Pat. 3,240,806 (1966).
318. Badische Anilin und Sodafabrik A.-G., Fr. Pat., 1,541, 629 (1968).
319. Distillers Co. Ltd., Belg. Pat., 659,233 (1965).
320. Eden, J. S., Brit. Pat., 1,131,162 (1968).
321. B. F. Goodrich Co., Neth. Appl., 6,612,181 (1967).
322. W. R. Grace and Co., Brit. Pat., 1,115,116 (1968).
323. Hirota, K., Kashiwara, H. and Nakamura, Y., Jap., 69 27,564 (1969).
324. I.C.I. Ltd., Neth. Appl., 6,513,697 (1966).
325. Japan Catalytic Chem. Industry Co. Ltd., Fr. Pat., 1,425,871 (1966).
326. Nakano, M., Komuro, I., and Koshikawa, T., Jap., 70 05,247 (1970).
327. Kashiwara, H., and Nakamura, Y., Jap., 68 24,645 (1968).
328. Eden, J. S., Fr. Pat., 1,554,521 (1969).

329. Trapasso, L. E., and Wenrick, J. D., U.S. Pat., 2,497,553 (1970).
330. Rohm and Haas Co., Neth. Appl., 6,604,795 (1966).
331. Scherhag, B., Hausweiler, A., and Schwartzer, K., Ger. Pat., 1,204,659 (1965).
332. Societe Nationale des Petroles d'Aquitaine, Fr. Pat., 1,533,971 (1968).
333. Young, H. S., and Reynolds, J. W., Ger. Offen., 1,811,541 (1969).
334. B. F. Goodrich Co., Neth. Appl., 6,500,418 (1965).
335. Parthasarthy, R., Dobres, R. M., and Warthen, J. L., U.S. Pat., 3,624,146 (1971).
336. Jakubowicz, L., Beres, J., Hauke, H., and Rataj, A., Fr. Pat., 1,544,453 (1968).
337. Kashiwara, H., and Nakamura, Y., Jap., 70 37,306 (1970).
338. Ondrey, J. A., and Swift, H. E., U.S. Pat., 3,641, 138 (1972).
339. Ondrey, J. A., and Swift, H. E., U.S. Pat., 3,655,750 (1972).
340. Frank, C. E., and Murib, J. H., Ger. Offen., 2,162,866 (1972).
341. Sennewald, K., Hauser, H., Gehrmann, K., and Lork, W., Brit. Pat., 1,220,568 (1971).
342. Ukibashi, H., Ota, Y., and Koshima, G., Jap., 71 09,134 (1971).
343. Kashiwara, H., and Nakamura, Y., Jap., 71 07,578 (1971).
344. Ono, I., Iikuni, T., Mizoguchi, J., Jap., 72 14,204 (1972).
345. Jakubowicz, L., Beres, J., Hauke, H., and Rataj, A., Pol., 61,669 (1971).
346. Komuro, I., Kadowaki, Y., and Kishikawa, T., Jap., 72 22,813 (1972).
347. Young, H. S., Anderson, G. C., and McDaniel, E. L., Fr. Addn., 95,230 (1970).
348. Kholyavenko, K. M., Baryshevskaya, I. M., Chernukhina, N. A., and Rubanik, M. Ya., *Katal. Katal. Akad. Nauk Ukr. SSR. Resp. Mezhvedom*, **2**, 64 (1966); *Chem. Abstr.*, **66**, 64828w.
349. Baryshevskaya, I. M., and Kholyavenko, K. M., *Katal. Katal. Akad. Nauk Ukr. SSR, Resp. Mezhvedom*, **5**, 23 (1969); *Chem. Abstr.*, **72**, 42535v.
350. Kholyavenko, K. M., Baryshevskaya, I. M., Maksimova, N. A., and Lazukin, V. I., *Katal. Katal. Akad. Nauk Ukr. SSR, Resp. Mezhvedom*, **6**, 61 (1970); *Chem. Abstr,* **74**, 124507t.
351. Baryshevskaya, I. M., Kholyavenko, K. M., and Rubanik, M. Ya., *Ukr. Khim. Zh.*, **33**, 984 (1967).
352. Baryshevskaya, I. M., Kholyavenko, K. M., and Rubanik, M. Ya., *Ukr. Khim. Zh.*, **35**, 702, 805 (1969).
353. Baryshevskaya, I. M., Kholyavenko, K. M., and Rubanik, M. Ya., *Katal. Katal. Akad. Nauk Ukr. SSR, Resp. Mezhvedom*, **6**, 53 (1970); *Chem. Abstr.*, **74**, 124508n.
354. Suvorov, B. V., Sembaev, D. Kh., and Rafikov, S. R., *Dokl. Akad. Nauk SSSR*, **172**, 1096 (1967).
355. Campbell, W. E., McDaniel, E. L., Reece, W. H., Williams, J. E., and Young, H. S., *Ind. Engng. Chem., Prod. Res. and Dev.*, **9**, 525 (1970).
356. Stobaugh, R. B., Clark, McH. S. G., and Cimirand, G. D., *Hydrocarbon Proc.*, **50**, 109 (1971).
357. Perkowski, J., *Przem. Chem.*, **51**, 17 (1972). *Chem. Abstr.* **76**, 88059d.
358. Navalkhina, M. D., Makotinskii, V. Yu., and Kotlobai, A. P., *Khim. Prom. (Moscow)*, **48**, 871 (1972); *Chem. Abstr.*, **78**, 30224d.
359. Caporali, G., *Hydrocarbon Process.*, **51**, 144 (1972).
360. Sampat, B. G., *Indian Chem. J.*, **6**, 145 (1971).
361. Heath, A., *Chem. Engr. (New York)*, **79**, 80 (1972).
362. Idol, J. D., Jr., U.S. Pat., 2,904,580 (1959).
363. Callahan, J. L., Foreman, R. W., and Veatch, F., U.S. Pat. 3,044, 966 (1962).
364. Callahan, J. L., Szabo, J. J., and Gertisser, B., U.S. Pat., 3,186,955 (1966).
365. Chemy Ind. (18), 697 (1972).

366. Yamada, K., Umemura, S., Fukuda, K., K., Sawazi, M., Ohdan, K., and Hidaka, M., Ger. Offen., 2,034,396 (1971).
367. Beckham, R., and Li, T. P., Ger. Offen., 2,013,915 (1970).
368. Yamada, K., Umemura, S., Ohdan, K., Hidaka, M. and Nakamura, Y., Ger. Offen., 2,039,011 (1971).
369. Ube Industries Ltd., Brit. Pat., 1,222,946 (1971).
370. Ioka, A., Yomitaniyama, A., Hizume, T., Kubota, Y., Furuhashi, S., and Sakai, T., Jap., 70, 35,287 (1970).
371. Honda, M., Dozono, T., Hirakawa, K., Aoki, K., Sugita, N., Ger. Offen., 2,104,016 (1971).
372. Yoshino, T., Saito, S., and Sobukawa, B., Jap., 71 03,438 (1971).
373. Yamada, K., Umemura, S., and Ohdan, K., Ger. Offen., 2,035,980 (1972).
374. Yamada, K., Nagai, S., Ohdan, K., and Hidaka, M., U.S. Pat., 3,629,317 (1971).
375. Notari, B., and Fattore, V., Ger. Offen., 2,139,565 (1972).
376. Ohordnik, A., Sennewald, K., Erpenbach, H., and Vierling, H., Ger. Offen, 2,104,223 (1972).
377. Montecatini Edison S.p.A., Fr. Pat., 1,538,997 (1968).
378. Hausweiler, A., Beilstein, G., Mayer, A., and Paris, N., Ger. Offen., 2,019,966 (1971).
379. Fattore, V., and Notari, B., Ger. Offen., 2,165,978 (1972).
380. Fattore, V., and Notari, B., Ger. Offen., 2,203,439 (1972).
381. Reulet, P., Tellier, J., Blanc, J. H., Joergensen, K. B. and Bohlbro, H., Fr. Addn., 2,080,231 to Fr. Pat., 1,563,988.
382. Taylor, K. M., Ger. Offen., 2,163,319 (1972).
383. Grasselli, R. K., Miller, A. F., and Hardman, H. F., Ger. Offen., 2,147,480 (1972).
384. Fattore, V., and Notari, B., Ger. Offen., 2,117,351 (1971).
385. Yoshino, T., Saito, S., Sofugawa, Y., and Sasaki, T., Jap., 72 19,764 (1972).
386. Yoshino, T., Saito, S., Ishikura, J., Sasaki, T., and Sofugawa, K., Jap., 71 02,804 (1971).
387. Reulet, P., Tellier, J., Pfister, A., Blanc, J. H., Joergensen, K. B., and Bohlbro, H., Fr. Addn., 2,033,428 (1970).
388. Yamada, K., Nagai, S., Ohdan, K., and Hidaka, M., Jap., 71 15,489 (1971).
389. Taylor, K. M., Ger. Offen., 1,964,786 (1970).
390. Yoshino, T., Ishikura, J., Sasaki, Y., Saito, S., Sobukawa, M., Ger. Offen, 1,811,063 (1969).
391. Aykan, K., U.S. Pat., 3,360,331 (1967) and U.S. Pat. 3,362,784 (1968).
392. Grayson, P. W., Lovett, G. H., Watts, K. B., and Fontenot, M. M., Ger. Offen, 2,152,539 (1972).
393. Cathala, M., and Germain, J. E., *Bull. Soc. chim. Fr.*, 2167, 2174 (1971).
394. Dalin, M. A., Lobkina, V. V., Abaev, G. N., Serebryakov, B. R., and Plaksunova, S. L., *Dokl. Akad. Nauk SSSR*, **145**, 1058 (1962).
395. Pasquon, I., Trifiro, F., and Centola, P., *Chimica Ind.*, *Milano*, **49**, 1151 (1967).
396. Gel'bshtein, A. I., Kul'kova, N. V., Stroeva, S. S., Bakshi, Yu. M., and Lapidus, V. L., *Probl. Kinet. Katal. Akad. Nauk SSSR*, **11**, 146 (1966). *Chem. Abstr.*, **66**, 2021f.
397. Margolis, L. Ya. *In* "Proceedings of the Fourth International Congress on Catalysis, Moscow (1968)" (D. A. Kazanaskii, ed.), Akademiai Kaido, Budapest, 1971.
398. Ohdan, K., Umemura, S., and Yamada, K., *Kogyo Kagaku Zasshi*, **72**, 2368 (1969).
399. Dalin, M. A., Mangasaryan, N. A., Serebryakov, B. R., Mekhtieva, V. L.,

Portyanskii, A. E., and Mekhtiev, K. M., *Dokl. Akad. Nauk SSSR* (Eng. transl.), **200**, 624 (1971).

400. Cathala, M., and Germain, J. E., *Bull. Soc. chim. Fr.*, 4114 (1970).
401. Shelstad, K. A., and Chong, T. C., *Can. J. chem. Engng*, **47**, 597 (1969).
402. Seeboth, H., Rieche, A., and Wolf, H., *Brennst. Chem.*, **50**, 268 (1969).
403. Giordano, N., quoted by Schuit, see Ref. 404.
404. Schuit, G. C. A., *Mem. Soc. Roy. Sci. Liege*, 6ᵉ series, tome I, Fasc. 4, p. 227 (1971).
405. Mulik, I. Ya., Rubanik, M. Ya., and Belousov, V. M., *Katal. Katal. Akad. Nauk Ukr. SSR, Resp. Mezhvedom. Sb.*, **3**, 121 (1967). *Chem. Abstr.*, **68**, 86832m.
406. Manogue, W. H., U.S. Pat., 3,360,331 (1967).
407. Aykan, K., U.S. Pat., 3,361,519 (1968).
408. Arai, H., Iiada, H., and Kunugi, T., *J. Catal.*, **17**, 396 (1970).
409. Aykan, K., *J. Catal.*, **12**, 281 (1968).
410. Aykan, K., and Sleight, A. W., *J. Am. Ceram. Soc.*, **53**, 427 (1970).
411. Nozaki, F., and Okada, H., *Nippon Kagaku Kaishi* (5), 842 (1972); *Chem. Abstr.*, **77**, 39625x.
412. Ohdan, K., Nagai, S., and Yamada, K., *Kogyo Kagaku Zasshi*, **72**, 2497, 2499 (1969).
413. Ohdan, K., Umemura, S., and Yamada, K., *Kogyo Kagaku Zasshi*, **73**, 441 (1970).
414. Boreskov, G. K., Ven'yaminov, S. A., Dzis'ko, V. A., Tarasova, D. V., Dindoïn, V. M., Sazonova, N. N., Olen'kova, I. P., and Kefeli, L. M., *Kinet. Katal.*, **10**, 1109 (1969).
415. Boreskov, G. K., Ven'yaminov, S. A., and Pankrat'ev, Yu. D., *Dokl, Akad. Nauk SSSR* (Eng. transl.), **196**, 621 (1971).
416. Skalkina, L. V., Suzdalev, I. P., Kolchin, I. K., and Margolis, L. Ya., *Kinet. Katal.* **10**, 218 (1969).
417. Trifiro, F., Centola, P., Pasquon, I., and Jiru, P. *In* "Proceedings of the Fourth International Congress on Catalysis, Moscow, 1968" (D. A. Kazanaskii, ed.), Akademiai Kaido, Budapest, 1971.
418. Trifiro, F., Lambri, C., and Pasquon, I., *Chimica Ind., Milano*, **53**, 339 (1971).
419. Crozat, M., and Germain, J. E., *Bull. Soc. chim. Fr.* (9), 3526 (1972).
420. Kominami, N., Nakajima, H., Kimura, T., Chono, M., Tamura, N., and Sakurai, T. *In* "International Symposium on Catalytic Oxidation, Imperial College, London, July 1970", The Chemical Society, London, 1970.
421. Asahi Chemical Co., Br. Pats., 1,084,599; 1,127,355; 1,139,398 and 1,156,620.
422. Kominami, N., Nakajima, H., Kimura, T., Chono, M., and Sakurai, N., Jap., 71 16,731 (1971).
423. Kominiami, N., Nakajima, H., Kimura, T., Chono, M., Tamura, N., and Sakurai, T., *Bull. Japan Petrol. Inst.*, **13**, 109 (1971).
424. Nakajima, H., Kimura, T., Kominami, N., Miyata, S., and Kobayashi, T., *Kogyo Kagaku Zasshi*, **74**, 1529, 1544, 2269, 2272, 2440 (1971); *Chem. Abstr.*, **75**, 109540a, 109542c, 109543d; **76**, 50515m, 100093v.
425. Moro-oka, Y., Tan, S., and Ozaki, A., *J. Catal.*, **12**, 291 (1968).
426. Ozaki, A., and Moro-oka, Y., Ger. Offen., 1,805,355 (1969).
427. Moro-oka, Y., Takita, Y., Tan, S., and Ozaki, A., *Bull. chem. Soc. Japan*, **41**, 2820 (1968).
428. Moro-oka, Y., Takita, Y., and Ozaki, A., *J. Catal.*, **23**, 183 (1971).
429. Buiten, J., *J. Catal.*, **13**, 373 (1969).

430. Buiten, J., *J. Catal.*, **27**, 232 (1972).
431. Moro-oka, Y., Takita, T., and Ozaki, A., *Bull. chem. Soc. Japan*, **44**, 293 (1971).
432. Moro-oka, Y., Takita, Y., and Ozaki, A., *J. Catal.*, **27**, 177 (1972).
433. Takita, Y., Ozaki, A., and Moro-oka, Y., *J. Catal.*, **27**, 185 (1972).
434. Pralus, C., Figueras, F., and de Mourges, L., *C. r. hebd. Séanc. Acad. Sci., Paris*, **C271**, 20 (1970).
435. Bak, T., Haber, J., and Ziolkowski, J., *Bull. Acad. pol. Sci. Sér. Sci. chim.*, **19**, 489 (1971).
436. Gati, G., and Mandy, T., *Chem. Ztg.*, **95**, 864 (1971).
437. Trimm, D. L., and Doerr, L. A., *J. Catal.*, **23**, 49 (1971).
438. Ohdan, K., Ogawa, T., Umemura, S., and Yamada, K., *Kogyo Kagaku Zasshi*, **73**, 824 (1970).
439. Sakamoto, T., Egashira, M., and Seiyama, T., *J. Catal.*, **16**, 407 (1970).
440. Seiyama, T., Egashira, M., Sakamoto, T., and Aso, I., *J. Catal.*, **24**, 76 (1972).
441. Friedli, H. R., Hart, P. J., and Vrieland, G. E., *Am. Chem. Soc., Div. Petrol. Chem.*, Prepr. 14, C70 (1969).
442. Swift, H. E., Bozik, J. E., and Ondrey, J. A., *J. Catal.*, **21**, 212 (1971).
443. Dorogova, V. B. and Kaliberdo, L. M., *Zh. Fiz. Khim.*, **45**, 2890 (1971); *Chem. Abstr.*, **76**, 45471q.
444. Massoth, F. E., and Scarpiello, D. A., *J. Catal.*, **21**, 225 (1971).
445. Gargarin, S. G., Gorshkov, A. P., and Margolis, L. Ya., *Kinet. Katal.* (Eng. transl.), **12**, 1269 (1972).
446. Bailey, W. J., *High Polym.*, **24**, 757 (1971).
447. Kaneko, K., and Wado, S., *Sekiyu Gakkai Shi*, **15**, 97 (1972).
448. Hearne, G. W., and Furman, K. W., U.S. Pat., 2,991,320 (1961).
449. Woskow, M. Z., Colling, P. M., and Karkalita, O. C., U.S. Pat., 3,428,703 (1969).
450. Callahan, J. L., Gertisser, B., and Grasselli, R., U.S. Pat., 3,257, 474 (1966).
451. Nolan, G. J., U.S. Pat., 3,320,329 (1967); 3,446,869 (1969).
452. Dow Chemical Co., Neth. Appl., 6,605,005 (1966).
453. Akiyama, S., and Karatsu, T., Ger. Offen., 1,941,513 (1970); 454 (a) Ger. Offen., 1,941,514 (1970); 454 (b) Ger. Offen., 1,941,515 (1970).
455. Bajars, L., U.S. Pat., 3,308,197 (1967).
456. Boutry, P., Daumas, J. C., and Montarnal, R., Ger. Offen., 1,816,847 (1969).
457. Boutry, P., Montarnal, R., and Wrzyszcz, J., Fr. Pat., 1,556,972 (1972).
458. Gaspar, J. H., and Pasternak, I. S., U.S. Pat., 3,320,331 (1967).
459. Hwa, F. C. S., and Bajars, L., U.S. Pat., 3,324,195 (1967).
460. Bajars, L., U.S. Pat., 3,308,196 (1967).
461. Noddings, C. R., and Gates, R. G., U.S. Pat., 3,501,549 (1970).
462. Nolan, G., and Hogan, R. J., Fr. Pat. 1,579,722 (1969); (a) Nolan, G., Hogan, R. J., and Farha, F., U.S. Pat., 3,501,547 (1970).
463. Pitzer, E. W., Ger. Offen., 2,124,454 (1971).
464. Nippon Kagaku Co. Ltd., Fr. Pat., 2,066,134 (1971).
465. Stowe, R. A., and Martin, I. J., U.S. Pat., 3,641,180 (1972).
466. Stowe, R. A., Hanger, Z. C., and Roberts, R. W., U.S. Pat., 3,541,172 (1970) and Ger. Offen., 1,908,861 (1970).
467. Yamaguchi, G., Komatsu, S., and Fukumoto, T., Ger. Offen., 2,141,286 (1972).
468. Pitzer, E. W., U.S. Pat., 3,686,346 (1972).
469. Nolan, G. J., and Holm, V. C. F., U.S. Pat., 3,580,969 (1971).
470. Japanese Geon Co. Ltd., Brit. Pat., 1,286,081 (1972).

471. Hagiwara, K., Ger. Offen., 2,039,168 (1971).
472. Beuther, H. and Swift, H. E., S. African Pat., 69 08,823 (1970) and Brit. Pat., 1,243,925 (1971).
473. Croce, L. J., and Bajars, L., U.S. Pat., 3,607,966 (1971).
474. Croce, L. J., and Bajars, L., U.S. Pat., 3,666,687 (1972).
475. Croce, L. J., and Bajars, L., Ger. Offen., 2,045,854 (1971).
476. Manning, H. E., Woskow, M. Z., and Christmann, H. F., Ger. Offen., 2,109,648 (1971).
477. Dean, J. C., and Colling, P. M., Ger. Offen., 2,118,344 (1971).
478. Kehl, W. L., U.S. Pat., 3,577,354 (1971). See also U.S. Pats, 3,595,809 and 3,595,810 (1971).
479. Pitzer, E. W., Ind. Engng Chem., Prod. Res. and Dev., 11, 299 (1972).
480. Voge, H. H., and Morgan, C. Z., Ind. Engng Chem., Prod. Res. and Dev., 11, 454 (1972).
481. Boutry, P., Montarnal, R., and Wrzyszcz, J., J. Catal., 13, 75 (1969).
482. Shih, Chien-Chou, Hua Hsueh, (2), 62 (1968); Chem. Abstr., 71, 112, 312a.
483. Polataiko, R. I., Shapovalova, L. P., and Musienko, V. P., Neft. Gazov. Prom. (6), 38 (1968); Chem. Abstr. 70, 77204x.
484. Emel'yanova, E. N., and Chugunnikova, R. V., Khim. Prom. (Moscow) 45, 254 (1969); Chem. Abstr. 71, 29968y.
485. Skarchenko, V. K., Kruglikova, N. S., Kuz'michev, S. P., and Vasilenko, L. P., Ukr. Khim. Zh., 35, 618 (1969); Chem. Abstr., 71, 80577x.
486. Skarchenko, V. K., Kruglikova, N. S., Luk'yanenko, V. P., Tyuryaev, I. Ya., and Golubova, E. E., Neft. Gazov. Prom., (5), 46 (1966); Chem. Abstr., 64, 4922e.
487. Pasternak, I. S., and Vadekar, M., Can. J. chem. Engng 48, 216 (1970).
488. Adams, C. R. In "Proceedings of the Third International Congress on Catalysis Amsterdam 1964", (W. M. H. Sachtler, G. C. A. Schuit, and P. Zwietering, eds) Vol. I, p. 240, North-Holland Publishing Co., Amsterdam 1965.
489. Serebryakov, B. R., Khiteeva, D. M., and Dalin, M. A., Khim. Prom. (Moscow) 44, 816 (1968); Chem. Abstr., 70, 46606a.
490. Keizer, K., Batist, Ph. A. and Schuit, G. C. A., J. Catal., 15, 256 (1969).
491. Roginskii, S. Z., Yanovskii, M. I., and Zimin, R. A., Neftekh, 7., 166 (1967).
492. Sadovnikov, V. V., Margolis, L. Ya., Yanovskii, M. I., and Zagvoskin, V. N., Gazov. Khromatog. (11), 48 (1969).
493. Batist, Ph. A., Prettre, H. J., and Schuit, G. C. A., J. Catal., 15, 267 (1969).
494. Batist, Ph. A., Van der Heijden, P. C. M., and Schuit, G. C. A., J. Catal., 22, 411 (1971).
495. Roginskii, R. A., Zimin, S. Z., and Yanovskii, M. I., Metody Issled, Katal. Katal. Reakts., Akad. Nauk SSSR, Sib. Otd., Inst. Katal., 3, 279 (1965); Chem. Abstr., 67, 90218n.
496. Watanabe, T., Kawakami, T., and Echigoya, E., Kogyo Kagaku Zasshi, 74, 2281 (1971).
497. Watanabe, T., Kawakami, T., and Echigoya, E., Kogyo Kaguku Zasshi, 74, 681 (1971).
498. Tsailingol'd, A. L., Pilipenko, F. S., Stepanov, G. A., and Tyuryaev, I. Ya., Neftekh., 6, 367 (1966); Chem. Abstr., 65, 13484e.
499. Sachtler, W. M. H., and De Boer, N. H. In "Proceedings of the Third International Congress on Catalysis, Amsterdam 1964", (W. M. H. Sachtler, G. C. A. Schuit, and P. Zwietering eds), Vol. I, p. 252, North-Holland Publishing Co., Amsterdam, 1965.

500. Mars, P., and Van Krevelen, D. W., *Chem. Engng Sci. Suppl.*, **3**, 41 (1954).
501. Alkhazov, T. G., Belen'kii, M. S., Khiteeva, V. M., and Alekseeva, R. I. *In* "Proceedings of the Fourth International Congress on Catalysis, Moscow 1968" (D. A. Kazanaskii, ed.), Akademiai Kaido, Budapest, 1971.
502. Watanabe, T., and Echigoya, E., *Kogyo Kagaku Zasshi*, **74**, 40 (1971).
503. Erman, L. Ya., Gal'perin, E. L., Kolchin, I. K., Dobrzhanskii, G. F., and Chernyshev, K. S., *Zh. Neorg. Khim.*, **9**, 2174 (1964).
504. Erman, L. Ya., and Gal'perin, E. L., *Zh. Neorg. Khim.*, **11**, 122 (1966).
505. Beres, J., Brueckman, K., Haber, J., and Janas, J., *Bull. Acad. pol. Sci. Sér. Sci. chim.*, **20**, 813 (1972).
506. Annenkova, B., and Alkhazov, T. G., *Nefteperab. Neftekh. (Moscow)* (12), 21 (1969); *Chem. Abstr.*, **72**, 89667e.
507. Annenkova, B., Alkhazov, T. G., and Belen'kii, M. S., *Kinet. Katal.*, **10**, 1076 (1969).
508. Morita, Y., Nishikawa, E., and Kiguchi, S., *Mem. Sch. Sci. Engng Waseda Univ.*, No. 32, 55 (1968); *Chem. Abstr.*, **71**, 105715f.
509. Adzhamov, K. Yu., Alkhazov, T. G., Belen'kii, M. S., and Lisovskii, A. E., *Kinet. Katal.*, **9**, 1279 (1968).
510. Adzhamov, K. Yu., Lisovskii, A. E., Alkhazov, T. G., and Belen'kii, M. S., *Neftekh.*, **9**, 86 (1969).
511. Mal'yan, A. N., Bakshi, Yu. M., and Gel'bshtein A. I., *Kinet. Katal.*, **9**, 1266 (1968).
512. Basner, M. E., Slin,ko, M. G., Tsailingol'd, A. L., and Ishchuk, I. V., *Teor. Osn. Khim. Tekhnol.*, **1**, 808 (1967); *Chem. Abstr.* **68**, 95114f.
513. Ter-Sarkisov, B. G., Aliev, F. V., Abilov, A. G., Aliev, N. M., and Aliev, V. S., *Azerb. Neft. Khoz.*, **48**, 37 (1969); *Chem. Abstr.*, **72**, 2946y.
514. Bakshi, Yu. M., Gur'yanova, R. N., Danilova, N. K., and Gel'bshtein, A. I., *Neftekh.*, **9**, 81 (1969).
515. Sekushova, Kh. Z., Vartanov, A. A., Alkhazov, T. G., and Belen'kii, M. S., *Izv. Vyssh. Ucheb. Zaved. Khim. Khim. Tekhnol.*, **13**, 102 (1970); *Chem. Abstr.*, **73**, 3283n.
516. Tarasova, D. V., Olen'kova, I. P., Dzis'ko V. A., Tovstonog, V. V., and Karakchiev, L. G., *Kinet. Katal.* (Eng. transl.), **12**, 1367 (1971).
517. Boreskov, G. K., Ven'yaminov, S. A., Dzis'ko, V. A., Tarasova, D. V., Dindoin, V. M., Sazonova, N. N., Olen'kova, I. P., and Kefeli, L. M., *Kinet. Katal.* (Eng. transl.), **70**, 1350 (1969).
518. Boreskov, G. K., Ven'yaminov, S. A., and Shchukin, V. P., *Dokl. Akad. Nauk SSSR* **192**, 831 (1970).
519. Shchukin, V. P., Boreskov, G. K., Ven'yaminov, S. A., and Tarasova, D. V., *Kinet. Katal.* (Eng. transl.), **11**, 153 (1970).
520. Shchukin, V. P., Ven'yaminov, S. A., and Boreskov, G. K., *Kinet. Katal.*, (Eng. transl.), **11**, 1236 (1970).
521. Shchukin, V. P., Ven'yaminov, S. A., and Boreskov, G. K., *Kinet. Katal.* (Eng. transl.), **12**, 547 (1971).
522. Boreskov, G. K., Shchukin, V. P., and Ven'yaminov, S. A. *In* "Okislitel'noe Degidrirovanie Uglevodorodov, Mater. Vses. Otraslevogo Soveshch. 1969" (publ. 1970), p. 87, (V. S. Aliev ed.), "Elm": Baku, Azerb. SSR; *Chem. Abstr.*, **75**, 64993g.
523. Malakhov, V. V., and Abdikova, F. G., *Kinet. Katal.* (Eng. transl.) **13**, 168 (1972).
524. Ammonsov, A. D., Sazonov, L. A., and Ven'yaminov, S. A., *Kinet. Katal.* (Eng. transl.), **12**, 536 (1971).

525. Shchukin, V. P., and Ven'yaminov, S. A., *Kinet. Katal.* (Eng. transl.), **11**, 1431 (1970).
526. Stroeva, S. S., Gel'bshtein, A. I., Kul'kova, M. V., Shcheglova, G. G., and Ezhkova, Z. I., *Metody Issled. Katal. Katal. Reakts. Sb*, **2**, 425 (1965); *Chem. Abstr.*, **65**, 16815e.
527. Trimm, D. L., and Gabbay, D. S., *Trans. Faraday Soc.*, **67**, 2782 (1971).
528. Trifiro, F., and Pasquon, I., *Chimica Ind., Milano*, **52**, 228 (1970).
529. Trifiro, F., Villa, P. L., and Pasquon, I., *Chimica Ind., Milano*, **52**, 857 (1970).
530. Alieva, M. M., Adzhamov, K. Yu., Shikhalizade, G. M., and Vartanov, A. A., *Uch. Zap. Azerb. Inst. Nefti. Khim.*, **9**, 79 (1971); *Chem. Abstr.*, **77**, 156825y.
531. Nishikawa, E., Kiguchi, S., and Morita, Y., *Kogyo Kagaku Zasshi*, **72**, 1278 (1969).
532. Kehl, W. L., and Rennard, R. J., U.S. Pats., 3,450,787 and 3,450,788 (1969).
533. Cares, W. R., and Hightower, J. W., *J. Catal.*, **23**, 193 (1971).
534. Rennard, R. J., and Kehl, W. L., *J. Catal.*, **21**, 282 (1971).
535. Massoth, F. E., and Scarpiello, D. A., *J. Catal.*, **21**, 294 (1971).
536. Pines, H., and Goetschel, C. T., *J. org. Chem.*, **30**, 3530 (1965).
537. Okamoto, Y., Happel, J., and Koyama, H., *Bull. chem. Soc. Japan*, **40**, 2333 (1967).
538. Delgrange, J. C., and Blanchard, M., *Bull. Soc. chim. Fr.*, 1093 (1971).
539. Andrushkevich, M. M., Buyanov, R. A., Timoshenko, V. I., and Spivak, S. I., *Kinet. Katal.* (Eng. transl.), **11**, 1419 (1970).
540. Afanas'ev, A. D., and Buyanov, R. A., *Kinet. Katal.* (Eng. transl.), **12**, 581 (1971).
541. Clark, A., and Shutt, R. S., U.S. Pat., 2,383,711 (1945).
542. Hearne, G. W., and Adams, M. L. U.S. Pat., 2,451,485 (1948).
543. Popova, N. I., and Mil'man, F. A., *Kinet. Katal.* **6**, 944 (1965); *Neftekh.* **6**, 90 (1966).
544. Gorokhovatskii, Ya. B., *Katal. Katal., Akad. Nauk Ukr. SSR, Resp. Mezhvedom. Sb.*, 71 (1965); *Chem. Abstr.*, **64**, 4894f.
545. Radzhabli, S. B., Khiteeva, V. M., Alkhazov, T. G., and Belen'kii, M. S., *Azerb. Khim. Zh.* (4), 18 (1970); *Chem. Abstr.*, **75**, 19516t.
546. Radzhabli, S. B., Khiteeva, V. M., and Alkhazov, T. G., *Dokl. Akad. Nauk Azerb. SSR*, **27**, 32 (1971); *Chem. Abstr.*, **77**, 4620q.
547. Bergman, R. I., and Frisch, N. W., U.S. Pat., 3,293,268 (1966).
548. I.C.I. Ltd., Neth. Appl., 6,514,701 (1966).
549. Marshall, D., Brit. Pat., 1,070,642 (1967).
550. Gilbert, G. P., Brit. Pat., 1,155,176 (1969).
551. Matsuura, R., Terahata, T., Komae, K., Tazawa, S., Daizawa, S., Imai, H., Minoda, S., Miyajima, M., Akiyama, T., and Ito, M., Ger. Offen., 1,951,537 (1970).
552. (a) Kerr, R. O., U.S. Pat., 3,255,211 (1966); (b) U.S. Pat., 3,255,212 (1966); (c) U.S. Pat., 3,255,213; (d) U.S. Pat., 3,288,721 (1966); (e) Petro-Tex Chem. Corp., Brit. Pat., 1,095,223 (1967).
553. (a) Nonnenmacher, H., Appl, M., Witwer, A., and Haug, J., Belg. Pat., 666,519 (1965) and Ger. Pat., 1,443,452 (1970); (b) Engelbach, H., Stoessel, A., and Renauer, E., Ger. Offen., 2,030,201 (1971); (c) Friedrichsen, W., Poehler, G., and Göhre, O., Brit. Pat., 1,154,148 (1969) and Fr. Pat., 1,495,765 (1967).
554. Kerr, R. O., Brit. Pat., 1,088,696 (1967); Fr. Pat., 1,451,364 (1966); Belg. Pat., 666,273 (1966).
555. Ai, M., Harada, K., and Suzuki, S., *Kogyo Kagaku Zasshi*, **73**, 524 (1970).

556. Ai, M., *Bull. chem. Soc. Japan*, **43**, 3490 (1970); *Bull. chem. Soc. Japan*, **44**, 761 (1971).
557. Ai, M., Niikuni, T., and Suzuki, S., *Kogyo Kagaku Zasshi*, **73**, 165 (1970).
558. Ai, M., *Sekiyu Gakkai Shi*, **14**, 324 (1971); *Chem. Abstr.*, **75**, 11791v8.
559. Ai, M., and Ishihara, T., *Kogyo Kagaku Zasshi*, **73**, 2152 (1970).
560. Ai, M., *Kogyo Kagaku Zasshi*, **74**, 183 (1971).
561. Akimova, L. S., Serebryakov, B. R., and Kolchin, I. K., *Neftekh.*, **11**, 545 (1971).
562. Ostroushko, V. I., Kernos, Yu. D., and Ioffe, I. I., *Neftekh.*, **12**, 362 (1972); *Chem. Abstr.*, **77**, 100433d.
563. Ai, M., and Boutry, P., *Bull. Soc. chim. Fr.*, 2775 (1970).
564. Ai, M., Boutry, P., Montarnal, R., and Thomas, G., *Bull. Soc. chim. Fr.*, 2783 (1970).
565. Ostroushko, V. I., Kernos, Yu. D., Saratova, D., and Moldavskii, B. L., *Neftekh.*, **9**, 886 (1969).
566. Ai, M., and Suzuki, S., *J. Catal.*, **26**, 202 (1972).
567. Blanchard, M., and Delgrange, J. C., *C. r. hebd. Séanc. Acad. Sci.*, *Paris*, **C262**, 1231 (1966).
568. Blanchard, M., and Delgrange, J. C. *C. r. hebd. Séanc. Acad. Sci.*, *Paris*, **C266**, 5 (1968).
569. Blanchard, M., Delplace, H., and Delgrange, J. C., *C. r. hebd. Séanc. Acad. Sci.*, *Paris*, **C269**, 1016 (1969).
570. Delgrange, J. C., and Blanchard, M., *Bull. Soc. chim. Fr.*, 1328 (1969).
571. Jescai, L., *Periodica polytech. Chem. Eng.* (*Budapest*), **13**, 157 (1969); *Chem. Abstr.*, **72**, 66274k.
572. Blanchard, M., Longuet, G., Boreskov, G. K., Muzykantov, V. S., and Panov, G. I., *Bull. Soc. chim. Fr.*, 814 (1971).
573. Sadykhova, Kh. I., Shakhtakhtinskii, T. N., and Elchieva, Z. M., *Azerb. Khim. Zh.* (2), 68 (1967); *Chem. Abstr.*, **68**, 12177e.
574. Avietsov, A. K., Gel'bshtein, A. I., Bakshi, Yu. M., and Gur'yanova, R. N., *Neftekh.*, **9**, 249 (1969).
575. Boutry, P., Courty, P., Daumas, J. C., and Montarnal, R., *Bull. Soc. chim. Fr.*, 4050 (1968).
576. Slovetskaya, K. I., Brueva, T. R., Dmitriev, R. V., and Rubinshtein, A. M., *Kinet, Katal.* **8**, 229 (1967).
577. Agasiev, R. A., Shakhtakhtinskii, T. N., Sadykhova, Kh. I., and Knopf, L. A., *Azerb. Khim. Zh.* (3), 25 (1969); *Chem. Abstr.* **72**, 66341e.
578. Agasiev, R. A., Ioffe, I. I., Shakhtakhtinskii, T. N., Sadykhova Kh. I., and Alieva, K. Ya., *Azerb. Khim. Zh.* (4), 60 (1969); *Chem. Abstr.*, **72**, 78346y.
579. Agasiev, R. A., Ioffe, I. I., Shakhtakhtinskii, T. N., Knopf, L. A., Shik, G. L., and Alieva, K. Ya., *Azerb. Khim. Zh.* (5), 128 (1969); *Chem. Abstr.*, **73**, 24875x.
580. Kuz'michev, S. P., and Skarchenko, V. K., *Kinet. Katal.* (Eng. transl.) **11**, 652 (1970).
581. Brockhaus, R., *Chem. Ing. Tech.*, **36**, 1039 (1966).
582. Brockhaus, R., Ger. Pat., 1,279,011 (1968).
583. Chemische Werke Huels A.-G., Fr. Pat., 1,470,474 (1967).
584. Brockhaus, R., Ger. Pat., 1,269,119 (1968).
585. Chemische Werke Huels A.-G. Fr. Pat., 1,479,681 (1967).
586. Brockhaus, R., Ger. Pat., 1,271,104 (1968).
587. Krabetz, R., Ger. Offen., 2,040,455 (1972).
588. Mizukami, T., Akahane, T., and Fuchigame, Y., Ger. Offen., 2,026,744 (1971).

589. Nakajima, K., and Sato, T., Jap., 70 4,569 (1970).
590. Nakajima, K., and Sato, T., Jap., 70 41,571 (1970).
591. Ito, J., Nakajima, K., and Sato, T., Jap., 70 4,367 (1970).
592. Santangelo, N., Battiston, G., Pregaglia, G., Croci, M., and Cavaterra, E., Ger. Offen., 2,053,120 (1971).
593. Santangelo, N., and Battiston, G., Ger. Offen., 2,110,876 (1971).
594. Brockhaus, R., and Hoeckele, G., Ger. Offen., 2,016, 681 (1971).
595. Hachmann, K., Gaube, J., Brockhaus, R., and Langheim, F., Ger. Offen. 2,059,945 (1972).
596. Kaneko, K., and Koyama, T., Ger. Offen., 2,164,023 (1972).
597. Kruit, J. H., and De Pagter, R. C. M. F., Neth. Appl. 70 14,629 (1972).
598. Kruit, J. H., Neth. Appl., 70 14,630 (1972).
599. Kruit, J. H., and De Pagter, R. C. M. F., Ger. Offen., 2,149,752 (1972).
600. Yasui, A., and Minakami, T., Jap., 72 10,692 (1972).
601. Yasui, A., and Minakami, T., Jap., 72 10,693 (1972); Jap., 10,694 and 10,695 (1972).
602. Uchijima, K., and Oda, Y., *Asahi Gavasu Kenkyu Hokoku*, **18**, 11 (1968); *Chem. Abstr.*, **70**, 67334a.
603. Takayama, Y., Ikeda, T., and Sunaoka, R., Jap., 68 10,603 (1968).
604. Mitsubishi Rayon Co. Ltd., Jap., 70 19,485 (1970).
605. Narita, T., and Nisijima, Y., Jap., 69 14,928 (1969).
606. Masaki, S., and Kuwata, T., Jap., 69 29,045 (1969).
607. Etherington, Jr., R. W., U.S. Pat., 3,254,035 (1966).
608. Izawa, M., Ono, I., and Noguchi., J., Jap., 68 03,164 (1968).
609. Cahoy, R. P., and Coyne, D. M., U.S. Pat., 3,417,144 (1968); (a) Cahoy, R. P., and Coyne, D. M., U.S. Pat., 3,342,869 (1967); (b) U.S. Pat., 3,499,936 (1970); and U.S. Pat., 3,557,021 (1971).
610. Brill, W. F., and Besozzi, A. J., U.S. Pat., 3,271,459 (1966); (a) Besozzi, A. J., and Brill, W. F., U.S. Pat., 3,301,906 (1967).
611. B. F. Goodrich Co., Neth. Appl., 6,500,728 (1965).
612. Rohm u. Haas G.m.b.H., Fr. Pat., 1,554,240 (1969).
613. Nakayama, Y., Ogawa, Y., and Uchida, T., Jap., 68 13,393 (1968); (a) Takayama, Y., and Ikeda, T., Jap., 69 08,992 (1969); (b) Nakayama, Y., Ogawa, Y., and Uchida, T., Jap., 69 13,129 (1969).
614. Oda, Y., Uchijima, K., and Fumami, F., Jap., 69 13,488 (1969).
615. Narita, T., and Nishijima, Y., Jap., 70 10,325 (1970).
616. Hagiwara, K., Ger. Offen., 2,038,998 (1971).
617. Takenaka, S., Shimizu, H., and Yamamoto, K., Ger. Offen., 2,155,411 (1972).
618. Nakajima, K., and Sadamichi, M., Jap., 71 41,885 (1971).
619. Adams, C. R., *Ind., Engng Chem.*, **61**, 30 (1969).
620. Mann, R. S., and Rouleau, C., *Can. J. chem. Engng*, **43**, 178 (1965); *Ind. Engng Chem., Prod. Res. and Dev.*, **3**, 94 (1964).
621. Vovyanko, I. I., and Gorokhovatskii, Ya. B., *Katal. Katal. Akad. Nauk SSSR, Resp. Mezh. Sb.*, **5**, 18 (1969); *Chem. Abstr.*, **72**, 48003s.
622. Gorokhovatskii, Ya. B., *Katal. Katal. Akad. Nauk SSSR, Resp. Mezh. Sb.*, **1**, 71 (1965); *Chem. Abstr.*, **64**, 4894f.
623. Mann, R. S., and Yao, K. C., *Ind. Engng Chem., Prod. Res. and Dev.*, **6**, 263 (1967); *Adv. Chem. Ser.*, No. 76, 276 (1968).
624. Mann, R. S., and Yao, K. C., *Ind., Engng Chem., Prod. Res. and Dev.*, **8**, 331 (1969).
625. Mann, R. S., and Yao, K. C., *Ind. Engng Chem., Prod. Res. and Dev.*, **10**, 25 (1971).

626. Mann., R. S., Yao, K. C., and Dosi, M. K., *J. Appl. Chem. Biotechnol.*, **22**, 915 (1972).
627. Zhiznevskii, V. M., Tolopko, D. K., and Fedevich, E. V., *Katal. Katal. Akad. Nauk Ukr. SSR, Resp. Mezhvedom Sb.*, **3**, 65, 76 (1967); *Chem. Abstr.*, **68**, 48743a, 48983d.
628. Zhiznevskii, V. M., Fedevich, E. V., and Tolopko, D. K., *Zh. Prikl. Khim.* (*Leningrad*), **44**, 846 (1971).
629. Fedevich, E. V., Zhiznevskii, V. M., and Tolopko, D. K., *Zh. Prikl. Khim.* (*Leningrad*), **4**, 70 (1968); *Chem. Abstr.*, **70**, 86960t and *Ukr. Khim. Zh.*, **36**, 400 (1970); *Chem. Abstr.*, **73**, 55553v.
630. Zhiznevskii, V. M., Fedevich, E. V., Tolopko, D. K., and Sulima, I. M., *Kinet. Katal.* (Eng. transl.) **12**, 374 (1971).
631. Zhiznevskii, V. M., Fedevich, E. V., and Sulima, I. M., *Neftepererab. Neftekh.* (*Moscow*) (4), 41 (1970); *Chem. Abstr.*, **73**, 27205q.
632. Zhiznevskii, V. M., Fedevich, E. V., and Sulima, I. M., *Ukr. Khim. Zh.*, **38**, 529 (1972); *Chem. Abstr.*, **78**, 15402s.
633. Fedevich, E. V., Zhiznevskii, V. M., and Krivoruchko, G. S., *Khim. Khim. Tekhnol.*, No. 1, 66 (1970); *Chem. Abstr.*, **77**, 87708y.
634. Zhiznevskii, V. M., Fedevich, E. V., and Sulima, I. M., *Neftekh.*, **11**, 594 (1971).
635. Malinowski, M., *Przem. Chem.*, **48**, 152 (1968); *Chem. Abstr.*, **71**, 30004n.
636. Zhiznevskii, V. M., Klyuchovskii, A. I., Tolopko, D. K., and Fedevich, E. V., *Zh. Prikl. Khim.*, **39**, 2540 (1966); *Chem. Abstr.*, **66**, 54766k.
637. Kakinoki, H., Mizushina, F., Kanda, A., Mita, I., and Ishikawa, T., *Sekiyu Gakkai Shi*, **10**, 255 (1967); *Chem. Abstr.*, **68**, 108282m.
638. Kakinoki, H., Mizushina, F., Kanda, A., and Suzuki, S., *Sekiyu Gakkai Shi*, **10**, 568 (1967); *Chem. Abstr.*, **69**, 76215u.
639. Kakinoki, H., Kanda, A., and Suzuki, H., *Sekiyu Gakkai Shi*, **10**, 722 (1967); *Chem. Abstr.*, **68**, 108283n.
640. Moro-oka, Y., Morikawa, Y., and Ozaki, A., *J. Catal.*, **7**, 23 (1967).
641. Serban, S., Goidea, D., and Anghel, N., *Rev. Chim.* (*Bucharest*), **17**, 466 (1966).
642. Kurata, N., Ohara, T., and Oda, K., Brit. Pat., 1,035,147 (1966).
643. Ball, W. J., and Gasson, E. J., Brit. Pat., 1,111,440 (1968).
644. Hapworth, J. P., and Baldwin, F. B., *Ind. Engng Chem.*, **34**, 1301 (1942).
645. Rellage, J. M., and Van der Vie, G. J. *In* "Tr. Mezhdunar. Konf. Kauch. Rezine 1969", p. 531, (P. F. Badenkov, ed.), Khimiya, Moscow, 1971; *Chem. Abstr.*, **77**, 76387s.
646. Alarie, D., *in* "Tr. Mezhdunar, Konf. Kauch. Rezine 1969", p. 523, (P. F. Badenkov, ed.), Khimiya, Moscow, 1971; *Chem. Abstr.*, **77**, 76386r.
647. Miyamori, H., Tomita, T., and Kaida, J., Jap., 72 38,425 (1972).
648. Cichowski, R. S., U.S., Pat. 3,660,514 (1972); Davison, J. W., U.S. Pat., 3,660,513 (1972).
649. Kara, N., Brit. Pat., 1,262,805 (1972).
650. Furuoya, I., Ogino, K., Kamatani, Y., and Naito, K., U.S. Pat., 3,662,016 (1972).
651. Naito K., Wada, T., Ogino, K. Kamatani, Y., and Furuoya, I., Jap., 71 00,161 (1971).
652. Belen'kii, M. S., Alkhazov, T. G., Vartanov, A. A., and Alyarbekova, O. A., *Izv. Vyssh. Ucheb. Zaved. Neft. Gaz*, **9**, 48 (1966); *Chem. Abstr.*, **66**, 39547w.
653. Gorokhovatskii, Ya. B., Pyatnitskaya, A. I., Korol', A. N. Popova, E. N., and Rozhkova, E. V., *Kinet. Katal.*, **9**, 81 (1968).
654. Vovyanko, I. I., and Gorokhovatskii, Ya. B., *Katal. Katal. Akad. Nauk SSR, Resp. Mezhvedom Sb.* (5), 18 (1969).

655. Adams, C. R. *In* "Proceedings of the Third International Congress on Catalysis, Amsterdam 1964" (W. M. H. Sachtler, G. C. A. Schuit and P. Zwietering, eds), I, p. 240, North-Holland Publishing Co., Amsterdam 1965.
656. Cant, N. W., and Hall, W. K., *J. Catal.*, **27**, 70 (1972).
657. Butt, N. S., and Fish, A., *J. Catal.*, **5**, 494 (1966).
658. Bretton, R. H., Wan, S. W., and Dodge, B. F., *Ind. Engng Chem.*, **44**, 594 (1952).
659. Butt, N. S., Fish, A., and Saleeb, F. Z., *J. Catal.*, **5**, 508 (1966).
660. Kolobikhin, V. A. *In* "Okislitel'noe Degidrirovanie Uglevodorodov, Mater. Vses. Otraslerogo Soveshch. 1969", p. 17 (V. S. Aliev, ed.), "Elm": Baku, Azerb. SSR (1970); *Chem. Abstr.*, **75**, 23507h.
661. Sakovich, A. V., *Neft. Gaz. Ikh. Prod.*, p. 159 (1971); *Chem. Abstr.*, **78**, 15403t.
662. Adel'son, S. V., Ivanovskii, B. L., Nikonov, V. I., and Sakovich, A. V., *Prom. Sin. Kauch.*, *Nauch-Tekh. Sb.*, No. 7, 7 (1971); *Chem. Abstr.*, **77**, 152610b.
663. Skarchenko, V. K., Luk'yanenko, V. P., and Kruglikova, N. S., *Neftepererab. Neftekhim.* (*Kiev*), No. 4, 146 (1971); *Chem. Abstr.*, **75**, 140147b.
664. Stergilov, O. D., Arinich, I. M., and Dudukina, T. V. *In* "Okislitel'noe Degidrirovanie, Uglevodorodov, Mater. Vses. Otraslevogo Soveshch. 1969", p. 188, (V. S. Aliev, ed.), "Elm": Baku, Azerb. SSR (1970); *Chem. Abstr.*, **75**, 38632u.
665. Stergilov, O. D., Arinich, I. M., and Dudukina, T. V., *Izv. Akad. Nauk SSSR, Ser. Khim.* (12), 2833 (1971); *Chem. Abstr.*, **77**, 33859e.
666. Bean, R. M. (Sun Oil Co.), U.S. Pat., 3,663,630 (1972).
667. Petro-Tex Chem. Corp., U.S. Pat., 3,671,606 (1972); U.S. Pat. 3,649,560 (1972); Ger. Offen., 2,138, 048 (1972).
668. Spoerke, R. W. (Goodyear Tyre and Rubber Co.), Ger. Offen., 2,138,611 (1972).
669. Centre de Mediations Scientifiques et Techniques, Fr. Demande, 2,108,849 (1972).
670. Popova, N. I., Kabakova, B. V., Mil'man, F. A., Latyshev, V. P., Vermel, E. E., Zhdanova, K. P., and Poltavchenko, Yu. A., *Probl. Kinet. i. Katal.*, *Akad. Nauk SSSR*, **11**, 153 (1966); *Chem. Abstr.*, **66**, 2024d.
671. Gorokhovatskii, Ya. B., Korol', A. N., Pismennaya, M. V., Popova, E. N., and Pyatnitskaya, A. I., *Kinet. Katal.* 8 (1967), Sb. Div. USSR Acad. Sciences.
672. Gorokhovatskii, Ya. B., Rozhkova, E. V., and Pyatnitskaya, A. I., *Kinet. Katal.*, **9**, 332 (1968); *Chem. Abstr.*, **69**, 76532v.
673. Belen'kii, M. S., Alkhazov, T. G., Khiteeva, V. M., and Radzhabli, S. B. *In* "Okislitel'noe Degidrirovanie Uglevodorodov, Mater. Vses. Otraslevogo Soveshch. 1969", p. 147 (V. S. Aliev, ed.), "Elm": Baku, Azerb. SSR (1970); *Chem. Abstr.*, **75**, 38672g.
674. Aliev, R. R., Gagarin, S. G., Yanovskii, M. I., and Zhomov, A. K., *Neftekh.*, **8**, 851 (1968).
675. Adams, C. R., *Ind. Engng Chem.*, **61**, 30 (1969).
676. Watanabe, T., Kuwajima, H., and Echigoya, E., *Kogyo Kagaku Zasshi*, **74**, 44 (1971); *Chem. Abstr.*, **74**, 111342k.
677. Gusman, T. Ya., Serebryakov, B. R., and Dalin, M. A., *Azerb. Khim. Zh.* (5), 33 (1967); *Chem. Abstr.*, **69**, 76224w.
678. Gagarin, S. G., Aliev, R. R., Yanovskii, M. I., Kholdyakov, N. I., and Zhomov, A. K., *Teor. Eksp. Khim.*, **8**, 244 (1972); *Chem. Abstr.*, **77**, 87507g.
679. Nishikawa, E., Ueki, T., and Morita, Y., *Kogyo Kagaku Zasshi*, **73**, 1673 (1970).
680. Usov, Yu. N., Skvortsova, E. V., Vaistub, Kh. G., and Kuchkaeva, I. K., *Keftekh.*, **12**, 481 (1972); *Chem. Abstr.*, **78**, 15271y.
681. Vaistub, Kh. G., Berman, A. D., Skvortsova, E. V., Usov, Yu. N., and Yanovskii, M. I., *Neftekh.*, **11**, 495 (1971).

682. Usov, Yu. N., Skvortsova, E. V., and Vaistub, Kh. G. *In* "Okislitel'noe Degidrirovanie Uglevodorodov, Mater. Vses. Ostraslevogo Soveshch. 1969", p. 141 (V. S. Aliev, ed.), "Elm": Baku, Azerb. SSR (1970); *Chem. Abstr.*, **75**, 23499g.
683. Butt, N. S., and Fish, A., *J. Catal.*, **5**, 205 (1966).
684. Bond, G. C., *Discuss. Faraday Soc.*, **41**, 200 (1966).
685. Wolkenstein, Th., *Adv. Catal.*, **12**, 189 (1960).
686. Balandin, A. A., *Adv. Catal.*, **19**, 1 (1969).
687. Dowden, D. A., and Wells, D., *in* "Actes du 2ᵉ Congres International de Catalyse, Paris 1960", **2**, p. 1489, Editions Technip, Paris, 1961.
688. Dowden, D. A., Proc. Fourth Internat. Cong. Catal. (Moscow, 1968), p. 163, Akad. Kiado: Budapest 1971.
689. Balandin, A. A., and Vasyunina, N. A., *Dokl. Akad. Nauk SSSR*, **103**, 831 (1955); *Dokl. Akad. Nauk SSSR*, **105**, 981 (1955).
690. Balandin, A. A., and Isagulyants, G. V., *Dokl. Akad. Nauk SSSR*, **63**, 139, 261 (1948).
691. Rubinstein, A. M., and Pribytkova, N. A., *Izv. Akad. Nauk SSSR, Otd. Khim. Nauk*, p. 509 (1945).
692. Tolstopyatova, A. A., *Vestn. Mosk. Univ.* No. 3, 47 (1951).
693. Balandin, A. A., *Zh. Obshch. Khim.*, **16**, 793 (1946).
694. Clark, A., "The Theory of Adsorption and Catalysis", Academic Press, London, 1970.
695. Lee, V. J., and Mason, D. R. *In* "Proceedings of the Third International Congress on Catalysis (1964) (W. M. H. Sachtler, G. C. A. Schuit and P. Zwietering, eds), p. 556, North-Holland Publishing Company, Amsterdam 1965.
696. Lee, V. J., *J. Catal.*, **17**, 178 (1970).
697. Dixon, G. M., Nicholls, D., and Steiner, H. *In* "Proceedings of the Third International Congress on Catalysis (1964)", Vol. 2, p. 815, North-Holland Publishing Company, Amsterdam, 1965.
698. Harrison, D. L., Nicholls, D., and Steiner, H., *J. Catal.*, **7** 359 (1967).
699. Krylov, O. V., "Catalysis by Non-metals" (transl. M. F. Delleo, Jr., G. Dembinski, J. Happel, and A. H. Weiss), Academic Press, London, 1970.
700. Shelef, M., Otto, K. and Gandhi, H., *J. Catal.* **12**, 361 (1968).
701. Roginskii, S. Z. *In* "Proceedings of the Fourth International Congress on Catalysis (Moscow 1968)", p. 16, Akademiai Kiado, Budapest, 1971.
702. Novakova, J., *Catal. Rev.*, **4**, 77 (1970).
703. Parravano, G., *Catal. Rev.*, **4**, 53 (1970).
704. Winter, E. R. S., *J. chem. Soc.*, 1522 (1954); *J. chem. Soc.*, 3824 (1955).
705. Boreskov, G. K., *Discuss. Faraday Soc.*, **41**, 263 (1966).
706. Boreskov, G. K., *Kinet. Katal.*, **11**, 374 (1970).
707. Boreskov, G. K., *Probl. Kin. i Kat. Akad. Nauk SSSR*, **11**, 45 (1966); *Chem. Abstr.*, **65**, 14483a.
708. Sazonov, V. A., Popovskii, V. V., and Voreskov, G. K., *Kinet. Katal.*, **9**, 312 (1968); *Chem. Abstr.*, **69**, 30505k.
709. Andrushkevich, T. V., Boreskov, G. K., Popovskii, V. V., Muzykantov, V. S., Kimkhai, O. N., and Sazonov, V. A., *Kinet. Katal.*, **9**, 595 (1968); *Chem. Abstr.*, **69**, 70192n.
710. Haber, J., and Grzybowska, B., *J. Catal.*, **28**, 489 (1973).
711. Mishchenko, Y. A., Adamiya, T. V., and Gel'bshtein, A. I., *Kinet. Katal.*, **11**, 927 (1970).
712. Boreskov, G. K., *Adv. Catal.*, **15**, 285 (1964).
713. Fahrenfort, J., Van Reijen, L. L., and Sachtler, W. M. H. *In* "Proceedings of the

Symposium on the Mechanism of Heterogeneous Catalysis (Amsterdam 1959)", p. 23, Elsevier Publishing Co., 1960.

714. Moro-oka, Y., and Ozaki, A., *J. Catal.*, **5**, 116 (1966).
715. Simons, Th. G. J., Verheijen, E. J. M., Batist, Ph.A., and Schuit, G. C. A., *Adv. in Chemistry Series*, **76**, Vol. II, p. 261 (1968).
716. Cornaz, P. F., Van Hooff, J. H. C., Pluijm, F. J., and Schuit, G. C. A., *Discussions Faraday Soc.*, **41**, 290 (1966).
717. Niwa, M., and Murikami, Y., *J. Catal.*, **26**, 359 (1972).
718. Niwa, M., and Murikami, Y., *J. Catal.*, **27**, 26 (1972).
719. Cole, D. J., Cullis, C. F., and Hucknall, D. J., unpublished results.
720. Badische Anilin und Sodafabrik, Brit. Pat., 1,140,264 (1969).
721. Catalysts and Chemicals Inc., Brit. Pat., 1,214,945 (1970).
722. Simard, G. L., Steger, J. F., Arnott, R. J., and Siegel, L. A., *Ind. Engng Chem.*, **47**, 1424 (1955).
723. Vol'fson, V. Ya., Zhigailo, V., Totskaya, E. F., and Raksha, V. V., *Kinet. Katal.*, **6**, 162 (1965).
724. Vanhove, D., and Blanchard, M., *Bull. Soc. chim. Fr.* (9), 3291 (1971).
725. Zemann, J., *Heidelberg. Beitr. Mineral. Petrog.*, M.5., 139 (1956).
726. Erman, L. Ya., and Galperin, E. L., *Zh. Neorg. Khim.*, **13**, 927 (1968).
727. Mekhtiev, K. M., Gamidov, R. S., Mamedov, Kh. S., and Belov, N. V., *Dokl. Akad. Nauk SSSR* **162**, 563 (1965).

Author Index

Numbers in italics are those pages on which References are listed

181

Subject Index

Numbers in italics indicate the page on which the main reference is to be found

196